Springer Theses

Recognizing Outstanding Ph.D. Research

For further volumes:
http://www.springer.com/series/8790

Aims and Scope

The series "Springer Theses" brings together a selection of the very best Ph.D. theses from around the world and across the physical sciences. Nominated and endorsed by two recognized specialists, each published volume has been selected for its scientific excellence and the high impact of its contents for the pertinent field of research. For greater accessibility to non-specialists, the published versions include an extended introduction, as well as a foreword by the student's supervisor explaining the special relevance of the work for the field. As a whole, the series will provide a valuable resource both for newcomers to the research fields described, and for other scientists seeking detailed background information on special questions. Finally, it provides an accredited documentation of the valuable contributions made by today's younger generation of scientists.

Theses are accepted into the series by invited nomination only and must fulfill all of the following criteria

- They must be written in good English.
- The topic of should fall within the confines of Chemistry, Physics and related interdisciplinary fields such as Materials, Nanoscience, Chemical Engineering, Complex Systems and Biophysics.
- The work reported in the thesis must represent a significant scientific advance.
- If the thesis includes previously published material, permission to reproduce this must be gained from the respective copyright holder.
- They must have been examined and passed during the 12 months prior to nomination.
- Each thesis should include a foreword by the supervisor outlining the significance of its content.
- The theses should have a clearly defined structure including an introduction accessible to scientists not expert in that particular field.

Kaden Richard Alan Hazzard

Quantum Phase Transitions in Cold Atoms and Low Temperature Solids

Doctoral Thesis accepted by
Cornell University, Ithaca, USA

Springer

Author
Dr. Kaden Richard Alan Hazzard
JILA University of Colorado
Boulder
USA
e-mail: kaden.hazzard@gmail.com

Supervisor
Dr. Erich J. Mueller
Department of Physics
Cornell University
Ithaca, NY 14853
USA
e-mail: em256@cornell.edu

ISSN 2190-5053 e-ISSN 2190-5061

ISBN 978-1-4419-8178-3 e-ISBN 978-1-4419-8179-0

DOI 10.1007/978-1-4419-8179-0

Springer New York Dordrecht Heidelberg London

Cover design: eStudio Calamar, Berlin/Figueres

Printed on acid-free paper

Springer is part of Springer Science+Business Media (www.springer.com)

To my Mom and Dad

Supervisor's Foreword

Since its birth in 1995, the field of ultracold atoms has been one of the most exciting areas of modern physics. Like the equally dramatic area of nanoscience, it is a synthetic field where the researchers engineer a system which will display interesting, novel, or useful characteristics. As technical hurdles are passed, the available toolbox grows, and the cold atom experiments become more and more interesting to a wider group of researchers.

In this very ambitious thesis, Dr. Kaden Hazzard sets out to accomplish three tasks: giving an overview of the many-body physics of cold atoms, presenting a technical roadmap for finding several specific pieces of new and important physics in the next generation of cold atom experiments, and providing a detailed analysis of current experiments. The first task is predominantly dealt with in his broadly accessible introduction, while the second and third are intertwined in the rest of the thesis. This structure is natural since it is only by understanding the capabilities and limitations of current experiments that one can sensibly move forward.

A particularly compelling and original aspect of this work is Dr. Hazzard's discussion of "quantum criticality" in cold atoms. The majority of quantum systems we are familiar with have low energy descriptions in terms of a gas of weakly interacting "quasiparticle" excitations. Near a continuous zero temperature phase transition, a "quantum critical point", this description breaks down. Over the past 20 years, understanding how to describe this quantum critical regime has been a major focus of condensed matter physicists. Among their many successes, researchers developed a powerful scaling paradigm, predominantly based on a mapping of d-dimensional quantum systems onto d+1 dimensional classical systems. At finite temperature, the extra "imaginary time" dimension has finite extent, and the standard techniques of finite size scaling constrain the behavior of all thermodynamic functions near the quantum critical point. Dr. Hazzard takes this now well-established theory, and asks how cold atom experiments can see quantum criticality. Can cold atom experiments teach us something new about this interesting regime and make a broad impact on other areas of physics? His answer is a resounding "yes", and he gives a number of concrete examples, such as the structure of the Mott-metal crossover in the Fermi Hubbard model. He even argues that with proper analysis, data from current experiments has the potential to make such an impact.

Equally important is Dr. Hazzard's treatment of spectroscopic probes of cold atoms. As an expert on spectroscopy, Dr. Hazzard takes many of his examples from that domain. Starting from a pedagogical introduction to cold collision shifts in radio/microwave spectra, he shows how to completely model one of the most important experiments in this area, accounting for all technical details. He then pushes the boundary, giving a strong coupling "Random Phase Approximation" analysis of radio frequency spectra of cold bosons in optical lattice. He uses an equation of motion approach to developing this theory, which is both simple to understand, but is very general and powerful. Through this technique he makes a number of dramatic predictions about the spectrum of trapped cesium atoms. Later in the thesis he gives extensive analysis of experiments on atomic hydrogen adsorbed on a helium surface, or embedded in a solid molecular hydrogen matrix. These latter systems sit between the areas of atomic and condensed matter physics, and show the generality of his techniques.

Both novice graduate students, and experts in cold atoms will benefit from reading this thesis. Not only does it give broad pictures and motivations, but it introduces powerful techniques; providing technical details which cannot be found elsewhere. Great care was taken with how the results are presented. Dr. Hazzard's conversational style conveys the excitement of this field without blunting the sharpness of the arguments. I believe this will be an important resource for years to come.

Ithaca, May 2011 Erich J. Mueller

Acknowledgments

I start by thanking the people who first introduced me to the professional world of physics. Foremost is John Wilkins, who graciously entertained inquiries regarding his research from me—a senior in high school—and invited me already to participate in conferences and in his research at The Ohio State University. To this day, John's generosity, immense knowledge, indefatigable work ethic, and helpful advice resonate with me; he was my first role model as a physicist.

Also, the rest of the Wilkins group deserve recognition, especially Jeongnim Kim, Seongwon Lee, David Richie, Dallas Trinkle, Richard Hennig, Erik Welch, and Joe Gorse, in no particular order. Each of them spent many hours (days, weeks, ...) of their life discussing physics and life, and this was tremendously enlightening and enjoyable. Also at Ohio State, a number of other professors profoundly influenced my view of physics: Richard Furnstahl, Thomas Lemberger, Cirayam Japaprakash, and Samir Mathur stand out.

After my second year as an undergraduate, I spent the summer doing experiments at Los Alamos National Labs with Joe Thompson, who exposed me to a whole new side of physics. It was at this time that I developed some knowledge of experimental physics, strongly correlated systems, and (to my current officemates' chagrin) an interest in quantum criticality. Seeing the importance of this later topic to understand real materials and the deep theoretical issues it raises is what has motivated me to try to bring its study to cold atoms.

During my time at Cornell, wonderful colleagues have been constantly present. My advisor Erich Mueller has been an inspiration: his ability and desire to approach problems with a "whatever-works," rather than a "whatever-we-know-how to-do" or "what-is-fashionable" attitude stands out. So do his familiarity with both deep theoretical approaches and the dirty details of particular experiments. I certainly have learned more physics from him than any other single person. Moreover, he has allowed me a rare degree of freedom and flexibility, for which I am grateful.

Through conversations and lectures, the Cornell physics faculty has been as diverse and exciting as one could wish for. Also, a number of faculty at other institutions have been welcoming and willing to converse, collaborate, and entertain. I particularly enjoyed John Shumway and Nikolay Prokof'ev's sabbatical visits. Also, thanks to Cheng Chin and the rest of his group, especially Nate Gemelke, for being so generous with their experimental data. I wish to thank Cornell's condensed matter faculty—naming them all would quickly turn into a laundry list. However, two deserve special recognition: Dave Lee who, despite his eminence, befriends year after year of nameless graduate students, and who has shared countless wisdom with me; Cyrus Umrigar for serving on my committee; and Mukund Vengalattore for spending an immense amount of his time sharing his expertise and advice with me, and having a fearless willingness to allow me near his optics tables, not to mention bravely offering to let me bench press him.

A number of friends and companions enriched my life and made it more fun. Again, a comprehensive list would be impossible to construct, so I will name a few that have been particularly important to me. The Mueller group have been my closest companions: Dan "betes" Goldbaum, Sourish "Tandoori-in-my-house" Basu, Stefan "Lil 'Bau Wau" Baur, Stefan "Naughty" Natu (D2), and Eliot Kapit. Dan and the Stefans in particular: the amount of time we spent talking physics, eating (half of the time meals far out of our price range, compensating with the other half by eating barely edible leftovers found on the floors of various Collegetown institutions), watching football, playing Wii, making strangers cry at CTB, and making fun of each other is simply unfathomable. Dan, I especially want to thank you for that magical night in San Francisco.

Outside of the Mueller group, but staying in the physics department, thanks to Watson, Sumiran Pujari, JoJo Heinonen, Joern Kupferschmidt, Ben Machta, Phil Kidd, Dave Bernat, and about 50 unnamed others for putting up with me and generally being awesome. Frisbee Girl, as punishment for your resistance to letting me bench press you, I hope you are not offended by your demotion to a "friendly acknowledgement".

Thanks to a few more friends for keeping me company and challenging me in far ranging manners, Angela and Sophie in particular. Over most of the last year, thanks to Betty (aka Debbie Martin) for having been one of my best friends and always putting a smile on my face. Most importantly, thank you Anna for your "jokes" at CTB, nursing me through two bouts of flu, removing the cholera from my apartment, interesting talks, travels, skydiving, and so on.

I have a similar (and not always unrelated) affection for all the baristas in Collegetown, the Commons, the Mall, and downtown for keeping a steady stream of caffeine in my veins! You were great diversions from the onslaught of physics and physicists.

Thanks to my family: Caley and Bradley, you're my favorite niece and nephew. Caley, this thesis would have been incomplete without your brilliant and profoundly deep insights into quantum mechanics.

Thank you, especially, mom and dad for the phone calls, trips to Ithaca (and elsewhere), a place to sleep when I visit, and your support in general. I've had the fortune of always knowing that if anything happened, you would be there.

There are too many important people in my life to recognize them all. All who I have left off will be unsurprised by my incomprehensiveness—and should certainly accost me the next time we meet, so I can acknowledge them (possibly with food and drink) in person.

Finally, I thank the American taxpayer for funding this research. Erich Mueller's contribution was partially performed at the Aspen Center for Physics. This material is based upon work supported by Cornell's special graduate fellowship, the GAANN Award No. P200A030111 from the US Department of Education, by National Science Foundation grant Nos. PHY-0456261 and PHY-0758104, and by ARO Award W911NF-07-1-0464 with funds from the DARPA OLE Program.

Contents

Chapter 1
Introduction to Many-Body Physics in Ultracold Atomic Gases

1.1 Motivation: Many-Body Physics

> Today we cannot see whether Schroedinger's equation contains frogs, musical composers, or morality—or whether it does not. Richard Feynman

Many of the most important, interesting, and vexing problems we encounter are questions about the behavior of large collections of objects: How do atoms organize themselves to form diverse materials such as plastics, metals, fluids, and magnets? How do connected neurons collectively result in intelligence? How do competing organisms give rise to ecosystems? How do humans organize themselves into groups, economies, and societies? Each of these questions shares the common thread of being a question about how entities come together to give rise to behavior not obviously connected to the underlying entities. In physics, the study of large collections of interacting objects is referred to as "many-body physics."

There are several reasons that condensed matter and ultracold atomic physics are wonderful areas in which to study this emergence. Firstly, unlike many other fields, we have a good understanding of the constituent pieces—they obey Schroedinger's equation referenced above by Feynman—and we can specifically focus on how the individual pieces' behaviors turn into the collective behavior. Secondly, it is possible to do reproducible, tunable, quantitative experiments—in contrast, you can't repeatedly create human societies in test tubes. Remaining firmly grounded in experimental consequences is extremely useful when developing reliable theories for many-body systems, since these theories almost invariably involve uncontrolled or untested approximations. Cold atoms and condensed matter offer us a grounded way to develop new methodologies for understanding emergent behavior.

Example of emergence in many-body physics. For a concrete example, consider a gold atom, a tiny, basically spherical, transparent speck. None of metallic gold's

K. R. A. Hazzard, *Quantum Phase Transitions in Cold Atoms and Low Temperature Solids*, Springer Theses, DOI: 10.1007/978-1-4419-8179-0_1,

properties are readily apparent: its shine and color, its large heat capacity (think of a cold winter, when a metal doorknob feels much cooler than wood at the same temperature), and its low electrical resistance have no direct counterparts in the individual gold atom. Understanding how atoms collectively form metals was an early coup of solid state physics [1]. This theory gave us, for example, an understanding of why copper is a metal while diamonds are insulators (not to mention explaining semiconductors and resulting in the transistor). This leads to another fascinating observation: although the emergent metal bears little resemblance to the individual atoms, vastly different atoms can give rise to similar behavior. For example, all metals—be they aluminum, copper, silver, or gold atoms, or a combination—share most of gold's characteristics listed above. This is not a coincidence: understanding this universality culminated in the development of the renormalization group in the 1970s (see Refs. [2–7] for good overviews). These ideas and tools of emergence, universality, and the renormalization group have since permeated well outside of physics, for example to biology (flocking, swarming, and bacterial motion [8], population dynamics [9], and the theory of disease propagation [10–12]), chemistry (reaction/diffusion equations [13] and *ab initio* quantum simulations of molecules [14]), mathematics (bifurcations and period doubling [15] and extreme value distributions [16, 17]), civil engineering (traffic flow [18]), sociology (refinements and elaborations of sociology's "balance theory" [19]), and economics and finance (a wide variety of concepts, ideas, and models are provided in Refs. [20–22]). There are also some entertaining, popular overviews touching on this area: "Sync," by Strogatz [23], "The Collapse of Chaos," by Stewart and Cohen [24], and "Chaos" by James Gleick are particularly relevant [25]. These examples are illustrative—other examples abound.

Applications. Finally, a little more practically, the fruits of studying emergent phenomena in physics have frequently led to technological advances: the transistor, hard drive, laser, and MRI's (magnetic resonance images) are just a few examples which owe their existence to fundamental discoveries in condensed matter physics over the last century. As we presently confront many phases of matter which cannot be understood within our current framework of how properties emerge from their constituent particles [1], one imagines that these will have equally dramatic applications. One that is often talked about is the application of so-called topological states to quantum computing [26].

Cold atomic systems offer considerable tunability and control, frequently used for precision measurement—spectroscopy, accelerometry, magnetometry, tests of fundamental physics such as time-variations of fundamental constants or violations of the standard model, and much more [27, 28]. One can utilize this tunability to emulate real materials and models of materials, for example high temperature superconductors [29]. It is in the latter role as analog systems to explore many-body systems on which this thesis concentrates.

Quantum emulation. At a first glance, one might suspect that using cold atomic experiments to emulate condensed matter models is simply creating a glorified, and highly special purpose, computer. However, because we are interested in quantum mechanical systems, this conclusion is incorrect: classical computers are unable to efficiently simulate general quantum systems. To be precise, the resources required to simulate a general quantum system on a classical computer grow exponentially with the simulated system's size. In contrast, the resources required to emulate a system using cold atoms grow linearly with the system's size: to emulate twice as large of a system, one requires twice as large of an emulator!

A simple argument allows us to understand the difficulty in simulating quantum mechanics in classical systems. Consider a spin-1/2 lattice system for concreteness. If there are L sites, then there are 2^L basis states in the Hilbert space, requiring 2^L complex numbers to specify the amplitudes of a single wavefunction in this space—this leads to an exponential growth in computational resources just to *store* the wavefunction of the quantum system at a given instance of time. To give a feeling for the size of this, for $L = 10$, this requires a several gigabytes of storage, even with modest precision for each of the amplitudes; for $L = 400$, one would require storing more amplitudes than there are particles in the observable universe. Meanwhile, a few grams of material will have $L \sim 10^{23}$ lattice sites, each of which may possess many more degrees of freedom than a spin-1/2 system. Clearly, simulating quantum systems with any classical computer in this manner is infeasible. This is the case even though recent, thorough analysis from quantum information indicates that for the relevant observables of physical systems, this estimate is somewhat too pessimistic: the cost scales as the exponential of the surface area rather than volume of a system. This has led to the development of much improved classical simulation algorithms [30, 31], but the exponential cost almost certainly persists for general quantum systems.

A universal *quantum* computer would allow one to efficiently simulate quantum systems (efficiently here meaning, roughly, resources scaling polynomially in the size of the simulated system). An enormous effort is underway to realize such systems, but scalable quantum computers of power even remotely comparable to classical computers seem far off. In contrast, cold atoms realization of condensed matter models is well underway, with many novel results already found in these experiments—Sect. 1.3 gives a brief historical overview. Additionally, they provide embodiments of the physics that a quantum computer generally does not (e.g. mass flow), useful in potential applications.

Understood foundations and open frontiers in many-body physics. Although I argued that exactly calculating a general quantum system's properties is impossible, there are many cases that are tractable: using a combination of approximations (e.g., perturbation theory), simple models (e.g., weakly interacting systems), and/or focusing on special observables (e.g., local static properties), one may make progress towards understanding broad classes of these systems. Increasing this understanding is a primary goal of many-body theory.

Physicists made tremendous progress in understanding a broad class of experimentally relevant materials that culminated in what I term "traditional many body theory." Early results were found shortly after the advent of quantum mechanics and were heavily developed in the 1940s–1970s. "Fermi liquids," "quasiparticles," and "symmetry breaking" are key concepts at the core of these results, which very briefly may be summarized as saying that frequently at low energies strongly interacting systems may be described in terms of a set of fictitious, weakly interacting "quasiparticles." Moreover, many of the systems we care about in nature possess such a description [1].

Naturally, modern many-body theory has a large focus on finding descriptions of systems where this picture breaks down. Perhaps the most studied examples in condensed matter are exotic phases of matter (e.g., spin liquids, topological phases, glasses) and other systems with exotic, non-quasiparticle behavior (e.g., non-Fermi liquids, quantum critical matter). Examples of both classes will appear in this thesis, and impacting these subjects using cold atoms will be its main focus. The effects of disorder, finite size ("mesoscopics"), and far from equilibrium physics are additional, somewhat conceptually different frontiers that offer exciting opportunities for study in condensed matter and cold atoms [32]. Although condensed matter emulations are the focus of this thesis, cold atoms can also realize conditions similar to neutron matter (BEC/BCS crossover near unitarity [33]), and realize qualitative analogs of high energy nuclear phenomena (color superconductivity), which also impacts astrophysical phenomena (vortices in multicomponent superfluids, as argued to be found in neutron stars) [34–36].

1.2 Motivation: General Atomic Physics

Advances in atomic physics have led to an ability to measure the internal structure of isolated atoms with high precision. To give an example of the extraordinary precision that has been possible, the electron g factor has been determined to 13 significant digits and with theoretical input from quantum electrodynamics determines the fine structure constant α to ten digits [37]. Additionally, this ability to address the internal structure has led to unrivaled control of quantum systems.

Atoms possess several internal degrees of freedom: electronic charge, electronic spin, and nuclear spin excitations. Molecules additionally possess vibrational and rotational degrees of freedom. Electromagnetic radiation is frequently used to excite and probe these degrees of freedom, and lasers are one of the most powerful tools available. This is largely because they are monochromatic, coherent, and possess high intensity. The field studying atomic properties with these tools is termed "atomic and molecular optics."

Experiments with these tools provide information about fundamental physics—precision measurement of fundamental constants—and also provide stringent tests of quantum mechanics and the standard model by being able to test for small deviations from expectations and to directly work with single photons of light coupled to matter [38]. They provide the most accurate clocks, with frequencies having relative systematic uncertainties of only 10^{-15} (see Ref. [39] and references within), and this number is rapidly improving. In astrophysics, precision knowledge of spectra allow us to characterize composition of extraterrestrial objects: planets, interstellar gas, and stars. Indeed, many fields of astronomical observation are characterized by the region of the electromagnetic spectrum probed: e.g., "radio astronomy," "infrared astronomy," "optical astronomy," etc. Such methods even play a crucial role in the prospect for detecting extraterrestrial life by looking for life's characteristic building blocks [40]!

The tools also offer a number of applications beyond physics. In chemistry, spectroscopy has improved our knowledge of chemical compounds and their reactions; specific examples are numerous, but one timely example is environmentally important pollutants and greenhouse gases [41, 42]. This has led to a new type of chemistry, with reactions catalyzed and otherwise controlled of these reactions with external tools, such as lasers [43, 44] (one might even consider the laser light as a "reactant"). Extending these methods to more complex molecules has allowed study of biological molecules and processes: photosynthesis, signal transport in nerves and neurons, and energy production in cells, to name a few [45]. Perhaps the most publicly visible applications are to technology and entertainment. To give just a few examples, lasers are used to read and write high density optical storage for computers; for high bandwidth data transmission; to mark, weld, and cut metals and other materials; and for eye surgery and clean scalpels [46].

A new frontier in atomic physics is to focus on what happens when atoms are not completely isolated from each other, and even when they strongly influence each other. In particular, in cold atoms, this leads to a highly unique place to study many-body physics, discussed in the rest of this thesis.

1.3 History and Introduction to Many-Body Physics in Cold Atomic Systems

1.3.1 Background and Achievements

To improve control and measurement of atomic systems, two important components are reducing fluctuations by decreasing the temperature and reducing the complexities of interactions with other atoms by decreasing the atomic density. A typical modern cold atoms experiment may achieve temperatures as low as a few nanokelvin at densities of $n \sim 10 \mu m^{-3}$. Using these qualities together with the simplicity of the constituents, one can ensure that this quantum system is closed—with essentially no unaccounted for external degrees of freedom on the timescales of interest—and apply external potentials, for example lattices. The cold, dilute regime possesses favorable features for the study of quantum many-body systems. The low temperatures enable quantum effects to manifest at the incredibly low densities in these experiments; the diluteness of the gas compared to the effective range of the interaction ensures that the interactions have a simple description, frequently in terms of a delta function contact interaction. Despite diluteness in this sense, the interaction *energy* may be made extremely large compared to the other energy scales, opening the way to many-body physics [32, 47].

Historically, the first step into many-body physics in these systems was the achievement of Bose–Einstein condensation in dilute Bose gases in 1995, the culmination of decades of development of laser cooling, trapping, and other techniques. The initial focus in the late 1990s was on the study of weakly interacting gases for which $na^3 \ll 1$, with n the density and a the s-wave scattering length (see Chap. 2 for a definition of this quantity, a measure of the interaction strength). This is the realization of a model system introduced in the 1950s to understand superfluidity in a setting that was much more tractable than in the long-studied superfluid ^{4}He. Cold atoms physicists explored numerous phenomena in these gases: the equation of state, collective modes, vortices, and dynamics [47]. Quantum turbulence [48, 49] and reduced dimensionality [50] remain exciting examples of physics studied in these systems.

The field has continued realizing model systems of interest in condensed matter, but since roughly 2001, a large component focuses on models which are *strongly interacting* and possess open questions for modern condensed matter research: the Mott insulator/superfluid phase transition for bosons in optical lattices, the BEC/BCS crossover in strongly interacting fermions, disordered systems, and reduced dimensional systems [32]. To give examples of the relevance of these systems, the BEC/BCS crossover probes the simplest model of superconductivity/superfluidity going beyond the Bardeen–Cooper–Schrieffer (BCS) theory that describes conventional superconductors. Going away from the BCS limit, the pairs become more tightly bound. Eventually, even in the absence of a Fermi sea they are bound, and for strong enough binding are described as a BEC of pairs. The system in between these limits is similar to systems in nuclear physics [33], and it is worth noting that, similar to the unitarity limit of the BEC/BCS crossover, the pair size in the cuprate high-temperature superconductors is also comparable to the interparticle spacing [51, 52]. Cold atoms experiments have contributed to quantitative studies of the thermodynamics [33], dynamics[53], and effects of reduced dimensionality and polarization in these systems, which can lead to exotic states such as FFLO [54].

The frontier of research goals for the next few years is attempting to realize phenomena that are still more exotic, more relevant to condensed matter, and less understood: nematics in dipolar gases [55], the phase diagram of the Fermi Hubbard model, in particular the question of whether it displays d-wave superconductivity [29], rotation or other techniques to realize topologically ordered states such as fractional quantum Hall states and topological insulators (see Chap. 8), and exotic "spin liquid" states in frustrated antiferromagnets. Thanks to the ideas presented in Chaps. 13 and 14, I am hopeful that we may add a major category to areas impacted by cold atoms in the near future: quantum criticality.

Chapter 2 will provide a more complete introduction to the experimental and theoretical tools used in optical lattice experiments. Optical lattices are probably the most common experimental tool with which to explore exotic physics with the most direct connection to condensed matter models.

1.4 Challenges and Outline of Thesis

I briefly outlined a subset of the most important achievements of cold atoms towards the study of many-body physics. The community is also facing major challenges in moving forward to more exotic states. The most important few are probably:

1. *Cooling*. Although cold atoms systems are the coldest systems in the universe (with a record low temperature of 50 pK in one setup, and a few nK more typically), there is a constant effort to further decrease temperatures. In the context of many-body physics, the most important impetus is to push to temperatures well below the bandwidth of particles in deep optical lattices ($\sim$ nK in typical setups) and to temperatures where superexchange and magnetism in optical lattices becomes important ($\lesssim$100 pK) [56]. New cooling methods are needed, and many are being pursued.
2. *Equilibration*. Much of what we hope to learn from cold atoms experiments concerns many-body systems near equilibrium (although far-from-equilibrium processes are exciting as well). However, the dynamics of cold atomic systems slows down at low temperatures, and optical lattices further slow the motional degrees of freedom. Indeed, the time to quantum mechanically tunnel between lattice sites can be tens of milliseconds. A typical cloud lifetime is at most a few seconds, so this limits the time for the system to equilibrate. Indeed, the experimental consequences of this have recently become quite ubiquitous, and as temperatures are lowered—and correspondingly timescales of interest get longer—even more work will need to be done to deal with this separation of timescales. Loss-induced correlations may offer one route to move forward (Ref. [57] is one example).
3. *Hamiltonian engineering and state control*. In order to quantitatively emulate model systems of interest, one needs to be able to reliably and accurately create the relevant Hamiltonians. While experimentalists are presently able to do this for many cases of interest—including the Bose and Fermi Hubbard models—the accuracy is modest (errors on the order of a percent in the best cases) and frequently require very low energy scales. Finding ways of characterizing the discrepancies from simple models, improving the accuracy of the Hamiltonians, and extending this to allow higher energy scales (to alleviate the preceding two challenges) are important goals for the field. Related to this, better state control may allow creation of novel and interesting many-body correlated states without them being equilibrium states of a Hamiltonian, or faster more accurate preparation of equilibrium (or metastable) states of a Hamiltonian (effectively a cooling procedure).

4. *Probes and methods to characterize behavior.* Even once novel states and behavior are achieved in a cold atoms system, we require new tools to extract an understanding of the physics of these systems. A standard technique is absorption imaging, which gives the column integrated density of the cloud, but we are still using only a fraction of the information in these images. Moreover, we would like to understand response functions, correlations, and dynamics, which are inaccessible by straightforward density images.

The work in this thesis largely concentrates on the last two challenges. It describes how to create and probe novel phases of matter and exotic (non-quasi-particle) behavior in cold atomic gases. It focuses on situations whose physics is relevant to condensed matter systems, and where open questions about these latter systems can be addressed. It also attempts to better understand several experimental anomalies in condensed matter systems.

The thesis is divided into five parts. The first section or chapter of each part gives an introduction to the motivation and background for the physics of that part; the last section or chapter gives an outlook for future studies. Parts 1–4 (Chaps. 1–14) introduce and show different facets of how to learn about novel physics relevant to condensed matter using cold atomic systems. Part 5 (Chaps. 15–18) constrains explanations of several ill-understood phenomena occurring in low-temperature quantum solids and condensed matter systems and attempts to construct mechanisms for their behavior.

Part 1 (Chaps. 1 and 2) motivates and introduces condensed matter, cold atoms, and many-body physics. Part 2 (Chaps. 3–7) introduces optical lattices and describes ways of spectroscopically probing many-body physics, especially dynamics, near quantum phase transitions in these systems. Part 3 (Chaps. 8–12) discusses the effects of rotation and how this, as well as alternative methods, can create exotic states. This leads us to study optical lattices where particles possess non-trivial correlations between particles even within a site in Chaps. 10 and 11. Part 4 (Chaps. 13 and 14) introduces another route to studying exotic physics in cold atoms: examining finite temperature behavior near second order quantum phase transitions. Non-quasiparticle behavior generically manifests in the "quantum critical regime" occurring near these transitions. This behavior has been hidden in previous analyses of data, but Chap. 13 introduces a set of tools that can extract universal quantum critical behavior from standard observables in cold atoms experiments. Chapter 13 discusses near-term opportunities to use these tools to impact fundamental, open questions in condensed matter physics. Part 5 (Chaps. 15–18) leaves cold atomic systems and introduces several anomalous or interesting experimental results from low-temperature and solid state physics, constrains possible explanations, and describes attempts to construct mechanisms for their behavior. Chapter 16 shows that collisional properties between quasi-two-dimensional spin-polarized hydrogen atoms are dramatically modified by the presence of a helium film, on which they are invariably adsorbed in present experiments. Chapter 17 constrains theories regarding recent observations on atomic hydrogen defects in molecular hydrogen quantum solids. Finally, Chap. 18

proposes a mechanism for supersolidity that can account for many of the experimental observations. While it leaves any questions open, it still contains an intriguing mechanism for coexistence of superfluidity and solidity that involves disorder.

References

1. P. Coleman, Ann. Henri Poincare **4**, 559 (2003)
2. N. Goldenfeld, *Lectures On Phase Transitions and the Renormalization Group*, illustrated edition edn. (Addison-Wesley, Reading, 1992)
3. A. Altland, B. Simons, *Condensed Matter Field Theory* (Cambridge University Press, Cambridge, 2006)
4. P.M. Chaikin, T.C. Lubensky, *Principles of Condensed Matter Physics*, 1st edn. (Cambridge University Press, Cambridge, 2000)
5. J. Cardy, *Scaling and Renormalization in Statistical Physics* (Cambridge University Press, Cambridge, 1996)
6. J. Zinn-Justin, *Quantum Field Theory and Critical Phenomena*, 4th edn. (Oxford University Press, New York, 2002)
7. A. Zee, *Quantum Field Theory in a Nutshell*, 2nd edn. (Princeton University Press, Princeton, 2010)
8. G.I. Menon, in *Rheology of Complex Fluids*, ed. by J. Murali Krishnan, P.B. Kumar (Springer, 2010). doi:10.1007/978-1-4419-6494-6, arxiv:1003.2032
9. P. Turchin, *Complex Population Dynamics: A Theoretical/Empirical Synthesis (MPB-35)* (Princeton University Press, Princeton, 2003)
10. M.J. Keeling, P. Rohani, *Modeling Infectious Diseases in Humans and Animals*, 1st edn. (Princeton University Press, Princeton, 2007)
11. R.M. Anderson, R.M. May, B. Anderson, *Infectious Diseases of Humans: Dynamics and Control* (Oxford University Press, New York, 1992)
12. O. Diekmann, J.A.P. Heesterbeek, *Mathematical Epidemiology of Infectious Diseases: Model Building Analysis and Interpretation*, 1st edn. (Wiley, New York, 2000)
13. D. Kondepudi, I. Prigogine, *Modern Thermodynamics: From Heat Engines to Dissipative Structures* (Wiley, Chichester, 1998)
14. G.K. Chan, J.J. Dorando, D. Ghosh, J. Hachmann, E. Neuscamman, H. Wang, T. Yanai, in *Frontiers in Quantum Systems in Chemistry and Physics, Progress in Theoretical Chemistry and Physics*, ed. by S. Wilson, P. Grout, J. Maruani, G. Delgado-Barrio, P. Piecuch (Springer, Dordrecht, 2008), pp. 49–65
15. M.J. Feigenbaum, J. Stat. Phys. **21**, 669 (1979)
16. S. Coles, *An Introduction to Statistical Modeling of Extreme Values*, 1st edn. (Springer, London, 2001)
17. M. Mézard, A. Montanari, *Information, Physics, and Computation* (Oxford University Press, New York, 2009)
18. B.S. Kerner, *Introduction to Modern Traffic Flow Theory and Control: The Long Road to Three-Phase Traffic Theory*, 1st edn. (Springer, London, 2009)
19. S.A. Marvel, S.H. Strogatz, J.M. Kleinberg, Phys. Rev. Lett. **103**, 198701 (2009)
20. P.W. Anderson, K. Arrow, D. Pines, *The Economy As an Evolving Complex System* (Westview Press, Boulder, 1988)
21. W.B. Arthur, S.N. Durlauf, *The Economy As an Evolving Complex System II* (Westview Press, Boulder, 1997)
22. L.E. Blume, S.N. Durlauf, *The Economy As an Evolving Complex System, III: Current Perspectives and Future Directions (Santa Fe Institute Studies on the Sciences of Complexity)*, illustrated edition edn. (Oxford University Press, New York, 2005)

23. S.H. Strogatz, *Sync: How Order Emerges From Chaos in the Universe, Nature, and Daily Life* (Hyperion, New York, 2004)
24. I. Stewart, J.S. Cohen, *The Collapse of Chaos: Discovering Simplicity in a Complex World* (Penguin Books Ltd, London, 2000)
25. J. Gleick, *Chaos: Making a New Science*, anv rep edn. (Penguin (Non-Classics), London, 2008)
26. A.Y. Kitaev, Ann. Phys. **303**, 2 (2003)
27. S.A. Diddams, D.J. Jones, J. Ye, S.T. Cundiff, J.L. Hall, J.K. Ranka, R.S. Windeler, R. Holzwarth, T. Udem, T.W. Hänsch, Phys. Rev. Lett. **84**, 5102 (2000)
28. M. Fischer, N. Kolachevsky, M. Zimmermann, R. Holzwarth, T. Udem, T.W. Hnsch, M. Abgrall, J. Grnert, I. Maksimovic, S. Bize, H. Marion, F.P.D. Santos, P. Lemonde, G. Santarelli, P. Laurent, A. Clairon, C. Salomon, M. Haas, U.D. Jentschura, C.H. Keitel, Phys. Rev. Lett. **92**, 230802 (2004)
29. A. Cho, Science **320**, 312 (2008)
30. Z.-C. Gu, M. Levin, X.-G. Wen, Phys. Rev. B **78**, 205116 (2008)
31. J.I. Cirac, F. Verstraete, J. Phys. A Math. Theor. **42**, 504004 (2009)
32. I. Bloch, J. Dalibard, W. Zwerger, Rev. Mod. Phys. **80**, 885 (2008)
33. S. Giorgini, L.P. Pitaevskii, S. Stringari, Rev. Mod. Phys. **80**, 1215 (2008)
34. K. Maeda, G. Baym, T. Hatsuda, Phys. Rev. Lett. **103**, 085301 (2009)
35. M.G. Alford, A. Schmitt, K. Rajagopal, T. Schäfer, Rev. Mod. Phys. **80**, 1455 (2008)
36. P. Braun-Munzinger, J. Wambach, Rev. Mod. Phys. **81**, 1031 (2009)
37. D. Hanneke, S. Fogwell, G. Gabrielse, Phys. Rev. Lett. **100**, 120801 (2008)
38. M.O. Scully, M.S. Zubairy, *Quantum Optics* (Cambridge University Press, Cambridge, 1997)
39. S. Blatt, A.D. Ludlow, G.K. Campbell, J.W. Thomsen, T. Zelevinsky, M.M. Boyd, J. Ye, X. Baillard, M. Fouch, R.L. Targat, A. Brusch, P. Lemonde, M. Takamoto, F. Hong, H. Katori, V.V. Flambaum, Phys. Rev. Lett. **100**, 140801 (2008)
40. B. Barbier, M. Gargaud, H. Martin, J. Reisse, de C. Duve, *Lectures in Astrobiology* (Springer Science & Business, London, 2006), pp. 556.
41. D.G. Andrews, *An Introduction to Atmospheric Physics*, 2nd edn. (Cambridge University Press, Cambridge, 2010)
42. G. Guyot, *Physics of the Environment and Climate* (Wiley, New York, 1998)
43. A. Assion, T. Baumert, M. Bergt, T. Brixner, B. Kiefer, V. Seyfried, M. Strehle, G. Gerber, Science **282**, 919 (1998)
44. W. Demtröder, *An Introduction to Atomic and Molecular Physics*, 1st edn. (Springer, London, 2005)
45. R. Cotterill, *Biophysics: An Introduction*, 1st edn. (Wiley, Chichester, 2002)
46. F.J. Duarte, *Tunable Laser Optics*, 1st edn. (CRC Press, Boca Raton, 2003)
47. C.J. Pethick, H. Smith, *Bose–Einstein Condensation in Dilute Gases* (Cambridge University Press, Cambridge, 2001)
48. N. Berloff, Physics **2**, 61 (2009)
49. E.A.L. Henn, J.A. Seman, G. Roati, K.M.F. Magalhes, V.S. Bagnato, Phys. Rev. Lett. **103**, 045301 (2009)
50. D. Petrov, D. Gangardt, G. Shlyapnikov, J. Phys. IV **116**, 5 (2004)
51. O. Fischer, M. Kugler, I. Maggio-Aprile, C. Berthod, C. Renner, Rev. Mod. Phys. **79**, 353 (2007)
52. C.H. Chen, S. Cheong, A.S. Cooper, Phys. Rev. Lett. **71**, 2461 (1993)
53. G. Veeravalli, E. Kuhnle, P. Dyke, C.J. Vale, Phys. Rev. Lett. **101**, 250403 (2008)
54. M.M. Parish, S.K. Baur, E.J. Mueller, D.A. Huse, Phys. Rev. Lett. **99**, 250403 (2007)
55. T. Lahaye, C. Menotti, L. Santos, M. Lewenstein, T. Pfau, Rep. Prog. Phys. **72**, 126401 (2009)
56. T.-L. Ho, Q. Zhou, Phys. Rev. Lett. **99**, 120404 (2007)
57. A.J. Daley, J.M. Taylor, S. Diehl, M. Baranov, P. Zoller, Phys. Rev. Lett. **102**, 040402 (2009)

Chapter 2
Theoretical and Experimental Techniques Used to Explore Many-Body Physics in Cold Atoms, Especially Optical Lattices

2.1 Experimental Techniques

2.1.1 Common Experimental Techniques for all Cold Atoms Experiments: Trapping and Cooling

The first stage of producing an ultracold atomic gas is to vaporize a solid (with an "oven" or other source) to create a hot gas, whose center of mass is then slowed sufficiently to be trapped. The slowing is frequently accomplished by a "Zeeman slower"[1]: a series of current carrying coils that generate magnetic fields, and which decrease in intensity over the length of the slower, to create a magnetic field gradient. Atoms used in this technique possess an electron spin degree of freedom so that the internal energy of the atom depends on the applied magnetic field and thus the magnetic field gradient introduces a potential energy gradient that slows the atoms [1, 2].

Magnetic trapping works with similar magnetic forces but with some subtleties because the naive setup of a quadrupolar magnetic field leads to a magnetic field zero at the trap center. Magnetic field zeroes cause spin states to become degenerate and thus atoms can transition between spin states as they move through magnetic field variations near the magnetic field zero, and—since the states transitioned to may be either untrapped or even repelled—the atoms will be lost. This is particularly problematic as the zero occurs at the trap center, where the particle density is highest. Moving the trap around to create a time-averaged potential or by using asymmetric traps [1, 2] somewhat obviates this problem. Optical traps are possible (using techniques analogous to those used to create periodic lattice potentials discussed in Sect. 2.1.2) and have advantages, including

[1] Stark decelerators—very roughly speaking replacing magnetic fields with electric fields—are another option when the particles being decelerated possess an electric dipole moment, often used with molecules

K. R. A. Hazzard, *Quantum Phase Transitions in Cold Atoms and Low Temperature Solids*, Springer Theses, DOI: 10.1007/978-1-4419-8179-0_2,

the ability to generate degenerate spin mixtures [1, 2], but require high laser intensity.

After atoms are trapped, they are cooled in multiple stages [2]. The first stage is generally Doppler cooling: a laser is red-detuned to an optical transition and the relativistic Doppler effect brings atoms moving towards the laser closer to resonance, therefore causing them to absorb more photons from the laser and slow down. The last cooling is evaporation, in which the cloud cools itself by losing the most rapidly moving atoms. These stages may be supplemented with other cooling schemes [1], leading to temperatures that have reached 50 picoKelvin.

Most commonly, experimentalists work with alkali atoms. These are convenient: their simple internal structure and single unpaired electron allow for straightforward optical and magnetic manipulation. Nevertheless, extensions of the techniques developed for alkalis have enabled BEC's of metastable He [3, 4] and several isotopes of Cr [5], Yb [6], Ca [7], and Sr [8, 9], as well as nearly degenerate Feshbach or ground state KRb molecules [10]. References [1, 2] contain more details and references.

It is worth noting that at these temperatures, the true equilibrium state is not a Bose–Einstein condensate, but a solid. However, on the timescales of interest, the system may frequently be described as an equilibrium state of some Hamiltonian. The BEC is the simplest example: the loss rates to form molecules and small solid clusters are negligibly small (on the order of seconds) compared to the time to equilibrate to a BEC (on the order of tens of milliseconds).

To understand what drives the pursuit of such extraordinarily cold temperatures, let's consider the transition temperature of a dilute gas, neglecting interactions. This result will also be useful later. Dimensional analysis get us most of the way: when the characteristic length scale of quantum mechanical motion grows to become on the order of the interparticle spacing, particles will lose their distinguishability and a normal gas will Bose condense. The length scale of quantum mechanical motion is the thermal de Broglie wavelength $\lambda_T = \sqrt{\frac{\hbar^2}{2mk_BT}}$ for particles with mass m, temperature T, and $\hbar$ and k_B Planck's and Boltzmann's constant respectively. For density n, the interparticle spacing is $r_s = n^{-1/3}$, so equating these lengths gives the transition temperature

$$T_c \sim \frac{\hbar^2 n^{2/3}}{k_B m}. \tag{2.1}$$

To understand this microscopically and obtain the prefactor, consider a gas with density of states $\nu(\epsilon) = C\epsilon^{\alpha}\Theta(\epsilon)$ for some constants C and α, and with Θ the Heaviside step function (the energy zero is set to be that of the lowest energy state). At sufficiently low temperatures this generically describes the density of states. For free particles of mass m in three dimensions, one has $\alpha = 1/2$, but for other α this also describes trapped gases and other dimensions. For bosons, the occupation of a state with energy ϵ in a grand canonical system with chemical potential μ is[2]

$$n(\mu, \epsilon) = \frac{1}{e^{(\epsilon-\mu)/T} - 1}. \tag{2.2}$$

The chemical potential must be less than the lowest energy state, or one obtains unphysical (negative) occupations. Then the maximal occupations and consequently maximum number of particles occurs for $\mu = 0$. Thus there is a maximum number of particles

$$\begin{aligned} N_{\max} &= \int_0^\infty d\epsilon \frac{C\epsilon^\alpha}{e^{\epsilon/T} - 1} \\ &= C\Gamma(\alpha+1)\zeta(\alpha+1)T^{\alpha+1} \end{aligned} \tag{2.3}$$

where $\zeta(\eta) \equiv \sum_{n=1}^{\infty} \frac{1}{n^\eta}$ is the Riemann zeta-function and $\Gamma(\eta)$ is the Euler gamma function. The last equality follows by Taylor expanding $(e^{\beta\epsilon} - 1)^{-1}$. If we put in more particles than $N_{\max}$, the occupation formula Eq. 2.2 fails and the lowest energy mode becomes macroscopically occupied with the excess particles, signifying a phase transition to the Bose–Einstein condensed state. Thus, for a given number of particles N, the transition occurs at

$$T_c = \left(\frac{N}{C\Gamma(\alpha+1)\zeta(\alpha+1)} \right)^{1/(\alpha+1)}. \tag{2.4}$$

Note that C is an extensive constant, so that the transition temperature depends on the number of particles only through the density, as expected for the homogeneous system. For free particles in three dimensions, we have $C = \frac{Vm^{3/2}}{\sqrt{2}\pi^2\hbar^3}$ and $\alpha = 1/2$, so $T_c = \left(\frac{n\sqrt{2}\pi^2\hbar^3}{m^{3/2}\Gamma(3/2)\zeta(3/2)}\right)^{2/3} = \frac{2\pi\hbar^2}{m}\left(\frac{n^2/3}{\zeta(3/2)}\right)^{2/3} \approx 3.3\frac{\hbar^2 n^{2/3}}{m}$.

With this formula, it is easy to understand the contrast of the $\sim$100 nK transition temperatures for BEC in these dilute gases with that in ^{4}He, in which superfluidity onsets at 2.2 K. The difference of several millions in transition temperatures is associated with the much lower density of the dilute gases.

Cooling to these low temperatures enables one to access not only bosonic superfluidity, but also fermionic degeneracy for Fermi gases. The characteristic temperature at which a Fermi surface develops is set by the same condition as the boson degeneracy—the thermal de Broglie wavelength is on the order of the

---

[2] I will go between setting $\hbar$ and k_B to one and showing them explicitly throughout this thesis.

interatomic spacing. In this case, however, there is no phase transition, just a crossover between two qualitatively distinct behaviors.

With interactions, these ultracold temperatures open up numerous opportunities to explore many-body physics in both fermionic and bosonic gases. A broad class of these involves lattices, the creation of which is the focus of the following section.

2.1.2 Optical Lattice Experimental Techniques

Here I discuss how lattice structures are realized experimentally in cold atoms. Optical lattices play a major role in exploring many-body physics in cold atoms. At the broadest level, they slow the atomic motion, increasing the importance of the interaction energy relative to the kinetic energy. They play a special role in emulating condensed matter models, since at the microscopic level solid state systems usually have some lattice structure.

Experimentally, one creates optical potentials using the AC stark effect of detuned laser light in which the cloud sits. For strong lasers (many photons in the lasing mode) it is appropriate to treat the atom–light interaction by treating the light as a classical oscillating electrical field

$$\mathbf{E}(\mathbf{r}, t) = \mathbf{E_0} e^{i\mathbf{k}\cdot\mathbf{r}} \cos(\omega t) \tag{2.5}$$

There is also an oscillating magnetic field, but this has a much smaller effect on the atoms. We assume the atom has only one optical transition near resonant with the laser frequency, and thus we can treat it as a two-level atom with ground and excited state energy levels $|g\rangle$ and $|e\rangle$ described by the Hamiltonian

$$H_a = E_g^{(0)}|g\rangle\langle g| + E_e^{(0)}|e\rangle\langle e|. \tag{2.6}$$

The transition frequency ω' is given by $\hbar\omega' = E_e^{(0)} - E_g^{(0)}$. The relationship $\delta = \omega - \omega'$ defines the detuning frequency δ of the laser beam from the optical transition. The atom–light interaction Hamiltonian is

$$H_{a-l} = \gamma E(\mathbf{r}, t)|e\rangle\langle g| + \mathbf{H.c.} \tag{2.7}$$

where γ is a constant and **H.c.** denotes the Hermitian conjugate. The dipole approximation is justified here because the wavelength of light is much larger than the atomic dimensions, and so γ is proportional to the dipole matrix element $\langle e|\mathbf{r}|g\rangle$. To simplify the time-dependence, we make the transformation to a new time-dependent basis with a new excited state, $|\tilde{e}\rangle$, defined by

$$|\tilde{e}\rangle = e^{i\omega t}|e\rangle. \tag{2.8}$$

In terms of the new states, the total Hamiltonian $H = H_a + H_{a-l}$ is[3]

$$H = E_g^{(0)}|g\rangle\langle g| + \left(E_e^{(0)} - \hbar\omega\right)|\tilde{e}\rangle\langle\tilde{e}| + \frac{E_0}{2}(e^{2i\omega t} + 1)|\tilde{e}\rangle\langle g| + \mathbf{H.c.} \qquad (2.12)$$

Within the rotating wave approximation, we can neglect the $e^{2i\omega t}$ term since it oscillates rapidly compared to the energy splitting of the $|\tilde{e}\rangle$ and $|g\rangle$ states: $E_e^{(0)} - \hbar\omega - E_g^{(0)}$, which is just $\hbar(\omega' - \omega) = \hbar\delta$. So as long as the detuning δ is small relative to the energy splitting, the physics is well described by the approximate Hamiltonian

$$H = E_g^{(0)}|g\rangle\langle g| + \left(E_e^{(0)} - \hbar\omega\right)|\tilde{e}\rangle\langle\tilde{e}| + \frac{E_0}{2}|\tilde{e}\rangle\langle g| + \mathbf{H.c.} \qquad (2.13)$$

This is a static two-level system, and the Hamiltonian is easily diagonalized. However, the important regime has detunings much less than $(\gamma E_0)^2$, and we may treat the atom–light interaction via second order perturbation theory to give the energy level shift for the internal ground state[4]

$$E_g - E_g^{(0)} = \frac{2\gamma^2 I}{c\epsilon_0\delta} \qquad (2.14)$$

with $I = c\epsilon_0 E_0^2/2$ the laser intensity. The key result is that potential energy is proportional to E_0^2 and hence the laser intensity. Thus, by shaping intensity profiles, one can shape potential surfaces. In addition to the applications here, this is

---

[3] Note that to get the new Hamiltonian governing the system in the transformed basis, we need to ensure that its solution give the same equations of motion (EOM) in the original basis as the original Hamiltonian (this is inequivalent to simply substituting the definition of the new basis states in the old Hamiltonian). The EOM of $|g\rangle$ is unchanged. The EOM for $|e\rangle$ is, from the original Hamiltonian,

$$i\partial_t|e\rangle = H|e\rangle = E_e^{(0)}|e\rangle + E_0\cos(\omega t)|g\rangle. \qquad (2.9)$$

Substituting the definition of $|\tilde{e}\rangle$, we have

$$\begin{aligned} i\partial_t e^{-i\omega t}|\tilde{e}\rangle &= H|\tilde{e}\rangle = E_e^{(0)}e^{-i\omega t}|\tilde{e}\rangle + E_0\cos(\omega t)|g\rangle \\ \Rightarrow i\left[e^{-i\omega t}\partial_t|\tilde{e}\rangle - i\omega e^{-i\omega t}|\tilde{e}\rangle\right] &+ \frac{E_0}{2}\left(e^{i\omega t} + e^{-i\omega t}\right)|g\rangle = E_e^{(0)}e^{-i\omega t}|\tilde{e}\rangle. \end{aligned} \qquad (2.10)$$

Thus

$$i\partial|\tilde{e}\rangle + \frac{E_0}{2}(e^{2i\omega t} + 1)|g\rangle = (E_e^{(0)} - \omega)|\tilde{e}\rangle, \qquad (2.11)$$

exactly the EOM obtained from the new Hamiltonian given in Eq. 2.12.

[4] Had we directly done an appropriate time-dependent perturbation theory, we could have done this directly and bypassed the rotating wave approximation—it would have emerged automatically.

the same idea behind optical tweezers applied to small systems, an experimental technique used extensively in chemical and biological systems [11].

Note that in deriving Eq. 2.14 I neglected any finite lifetime of the states, but this arises due to spontaneous and stimulated photon emission and photon absorption. One can model the decay as adding an imaginary part $i\Gamma/2$, with Γ the decay rate, to the energy of the E_e state, in which case the above analysis yields

$$E_g - E_g^{(0)} = \frac{2\gamma^2 \delta I}{c\epsilon_0 \left(\delta^2 + \Gamma^2\right)}. \tag{2.15}$$

Then the simple picture above is valid only when the detuning is large compared to Γ (proportional to the linewidth), in which case we were justified in neglecting the real absorption and emission.

Using the observation that spatial variations of light intensity give spatial potential energy variations for the atoms, one can create various potential energy surfaces. The most common optical potential is made by interfering two coherent, counterpropagating lasers. The standing wave interference pattern yields a sinusoidal lattice potential

$$V(\mathbf{r}) = \sum_{\alpha=x,y,z} V_{0,\alpha} \sin(\mathbf{k}_\alpha \cdot \mathbf{r}) \tag{2.16}$$

for a set of lattice intensitis $V_{0,\alpha}$ and wavevectors $\mathbf{k}_\alpha$. Doing this in three orthogonal directions, one forms a cubic sinusoidal lattice. The many-body physics of bosons and fermions in such a lattice is extremely rich and related to important models of solid state physics. This will be introduced in Sect. 2.2.3 and discussed throughout this thesis.

This technique can be employed to create other optical external potentials. For example, if the counterpropagating lasers are incoherent, no interference pattern forms and the light intensity is just the beam profile, creating a trapping potential. Such optical potentials are increasingly used, and are sometimes necessary. For example, to create spinor gases, one must eliminate magnetic fields that break the spin degeneracy [12]. Another example is the patterning of arbitrary potential landscapes—using masks and high quality, high numerical aperture optics—that is possible in some experimental setups [13].

2.1.3 Detection

The main tool used to probe cold atomic gases is absorption imaging. In this technique, one shines a laser on the cloud, and a CCD camera on the other side of the cloud measures the light transmitted. In other word, one measures the cloud's shadow. In the limit where multiple scattering can be neglected (valid for many experiments), the transmitted intensity I_t at position (x, y) is

$$I_t(x, y) = I_i e^{-\tau} \tag{2.17}$$

with I_i the incident intensity, and τ the optical depth is given by

$$\tau = N(x,y)\sigma \tag{2.18}$$

where $N(x,y)$ is the cross sectional density of atoms in the path of the beam captured at the pixel with location (x,y), and σ is the scattering cross section for the light scattering off of an individual atom. Thus, we see that measuring I_t/I_i allows one to determine the column integrated density of the cloud.

By applying various protocols before imaging, one can obtain much more general information than the density profile. Spectroscopy is one example—one can create excitations of the cloud and then image the system after this procedure. Chapters 3–7 discuss this in the context of radio-frequency spectroscopy. Another, very common protocol, is *time-of-flight* imaging. Here, one instantaneously turns off all external potentials and interactions and lets the cloud expand. After a sufficiently long time, the density at a position $\mathbf{r}$ is proportional to the initial momentum distribution at a velocity $\mathbf{k} \propto \mathbf{r}$: thus for a condensate, the real space image is just the momentum space structure factor. To see this, note that the sudden turn-off of the trap, lattice, interactions, etc. projects the quantum state onto momentum–space eigenstates $|\mathbf{k}_1, \mathbf{k}_2, \ldots, \mathbf{k}_N\rangle$ (for N-particles). The dynamic evolution then displaces a state with momentum $\mathbf{k}$ by $\hbar\mathbf{k}t/m$ in time t. For long times, the initial position is irrelevant and the final position is proportional to $\mathbf{k}$, as claimed.

The resolution of these imaging techniques varies, but in the past the imaging resolution has been much coarser than the atomic spacing[5] However, very recent developments enable imaging of single atoms with high resolution. At least six such experimental setups have been demonstrated. In 2007 David Weiss's group gave an early demonstration of imaging atoms in an optical lattice with single-site accuracy [14] by employing a large lattice spacing (several microns). Due to the large lattice spacing, however, the tunneling energy was negligible, so that there was no quantum mechanical coupling between the sites—all observed motion was thermal activation. In 2008, using an scanning electron microscope, Herwig Ott's group measured individual atoms in optical lattices with lattice spacings useful for doing many body physics, with resolution a fraction of the lattice spacing [15]. While this technique provides spectacular resolution for these gases, the detection fidelity is rather low: only $\sim 10\%$ of the atoms are detected, somewhat limiting the utility for, say, measuring density fluctuations or spatial correlations. In 2009, through the use of a two-dimensional system near the imaging optics and extremely high quality, high numerical aperture optics, Cheng Chin's group has measured [16] the density profiles and fluctuations of bosons in an optical lattice in the Bose–Hubbard regime with linear spatial resolution of 2–3 sites.[6] At the end of

[5] This is another reason for the prominence of time-of-flight imaging, since it expands the imaged cloud's size.

[6] The pixel size is slightly smaller. At the moment, other imaging elements limit the resolution, but may be removed if one can properly account for the point-spread function of the optics.

2009, Markus Greiner's group demonstrated single site resolution imaging (with near unity fidelity) of a BEC in an optical lattice [13], approaching the Mott insulator regime by using a two-dimensional gas near the imaging optics, high quality optics, and a large numerical aperture. The main drawback of the approach is that the imaging light induces atomic to pair into molecules when they are on the same site, and thus only measures the parity of the occupation—n *mod* 2—not the occupation n.

2.2 Theoretical Techniques

2.2.1 Interactions in Cold Atom Systems

Interactions are the key ingredient in many-body physics—without them, the behavior is reducible to single particle physics! In this section, I introduce the appropriate description of interactions in cold atoms experiments. Good (broader) discussions can also be found in the books [1, 17].

The first point to be emphasized is that the word "dilute" is used with (at least) two very distinct meanings. Both involve the interactions. The first is that the interaction energy is small—in three dimensions this requires that the scattering length a (defined shortly) satisfies $na^3 \ll 1$ with n the density.[7] Although many early cold atoms experiments were in this regime, the most interesting many-body physics requires strong interactions. Consequently many modern cold atom experiments explore regimes that are not dilute in this sense.[8]

Although experiments on many-body physics are not dilute in this first sense, there is a second sense in which virtually all cold atoms experiments are dilute: the *range* of the interaction, r_{eff} satisfies $nr_{\text{eff}}^3 \ll 1$. Note that r_{eff} is in principle independent of the scattering length. This second sense of diluteness allows one to describe the interactions with an appropriately regularized delta function, with second quantized Hamiltonian

---

7 This criterion follows from a simple argument: basically by dimensional analysis (assuming that the interaction energy is proportional to a, motivated later), the kinetic energy per particle is $E_k \sim n^{2/3}$ and the interaction energy per particle is $E_i \sim an$. Requiring $E_i \ll E_k$ gives $na^3 \ll 1$.

8 A warning about notation is in order. Ostensibly, this argument would preclude experiments from studying many-body physics in the Bose–Hubbard model, since these experiments are in a regime $na^3 \ll 1$—in fact, this is a necessary criterion for the applicability of the Bose–Hubbard model to describe optical lattice experiments (see Sect. 2.2.3). However, the "a" appearing in the text's criterion is the scattering length for two particle problem *in whatever external potential is applied*, including any optical lattice. The effective scattering length in a lattice—call it a_L—is much larger than the free space a. Indeed, for strong lattices is on the order of the lattice spacing d. Thus, the scattering length a_L describing two low energy particles in a lattice is comparable to the density in the Bose–Hubbard regime as one approaches the Mott insulator, and one is indeed in a strongly correlated regime.

$$H_{\text{int}} = \frac{g}{2} \int d\,\mathbf{r}\, \psi^\dagger(\mathbf{r})\psi^\dagger(\mathbf{r})\psi(\mathbf{r})\psi(\mathbf{r}) \tag{2.19}$$

where $\psi^\dagger(\mathbf{r})$ and $\psi(\mathbf{r})$ are creation and annihilation operators for particles at position $\mathbf{r}$, respectively. Some care is required in having the interactions occur only at a single point in space, and I will return to this momentarily.

Our goal is to determine the coupling constant g in Eq. 2.19. One can imagine different philosophies with which to approach this question. One can imagine computing the interaction energy for two particles in some configuration to determine g from an ab initio description of the atoms in terms of the nuclei and electrons comprising them. However, this is very challenging, and despite the fact that g is just a measure of two particles' interactions, it is still impossible to compute g with significant accuracy for many-electron atoms even with sophisticated numerical techniques and computational power. A more general philosophy is to try to relate g to independently accessible experimental measurements. Here, we will see that measuring two particle scattering properties allows one to determine g.

Our approach is to formally calculate the large distance scattering properties of the true interatomic potential and do the same for the contact interaction. We will see these can be matched with a proper choice of g, justifying our description of the system's physics in terms of a delta function, since the atoms are always separated by distances large compared to the potential's range. Define $V(r)$ to be the true two-particle interatomic interaction potential at interparticle separation r. The free space two-particle problem reduces to a single particle problem by working in relative coordinates where the wavefunction satisfies the Schrödinger equation [18]

$$-\frac{\hbar^2}{m}\nabla^2\psi(\mathbf{r}) + V(r)\psi(\mathbf{r}) = E\psi(\mathbf{r}) \tag{2.20}$$

for the three dimensional wavefunction ψ. Moreover, at the low energies of interest, s-wave scattering is the only relevant scattering channel, so we use the Schrödinger equation for the s-wave channel in radial coordinates [18]:

$$-\frac{\hbar^2}{mr}\frac{d^2}{dr^2}[rR(r)] + V(r)R(r) = \frac{\hbar^2k^2}{m}R(r) \tag{2.21}$$

where R is the s-wave radial wavefunction and k is the asymptotic momentum.

One can characterize the large distance scattering with the scattering length a. This is defined to be the radius of a hard sphere (yet another potential) that would give the same long distance scattering properties. Since we want to match long-distance wavefunction, I start by calculating the large distance, low energy wavefunction for the hard sphere potential. Equation 2.21 for a hard sphere of radius a is

$$-\frac{1}{r}\frac{d^2}{dr^2}[rR(r)] = k^2R(r) \quad \text{for r} > a \tag{2.22}$$

supplemented with a boundary condition $\psi(a) = 0$. (As a second order differential equation, one needs a second boundary condition, but this is equivalent to choosing a normalization for the wavefunction.) This equation is just $f''(r) = k^2 f$ for $f \equiv rR$, so the solution with $R(a) = 0$ is, for $r > a$ (up to normalization factor)

$$\begin{aligned} R(r) &\sim \frac{1}{r}\sin[k(r-a)] \\ &\sim 1 - \frac{a}{r} \end{aligned} \tag{2.23}$$

expanding for low energies—$kr, ka \ll 1$—in the second line.

Now let's examine the form of the large distance scattering problem. Choose the incident wave to come along the z axis; angular momentum channels decouple at large distances, so for potential decaying sufficiently rapidly (faster than some power) we can write the large distance wavefunction as

$$\psi(r,\theta) = e^{ikz} + f(\theta)\frac{e^{ikr}}{r} \tag{2.24}$$

with θ the angle of from the z axis and $f(\theta)$ called the scattering amplitude. At low energies and large distances, the s-wave channel dominates, the scattering wavefunction simplifies to $\psi \to 1 + f/r$. Comparing with Eq. 2.23 shows that scattering in *any* potential gives a large distance scattering wavefunction of the same form as that of the hard sphere's. Moreover, it shows how to relate the scattering length a describing an arbitrary potential to the experimentally observable two-particle scattering cross section σ, since[9] $\sigma = 8\pi a^2$.

There is one final step: we compute the scattering length corresponding to a contact interaction with strength g; this will tell us how to choose g to reproduce the scattering properties of an arbitrary potential with some measured scattering length. Some subtlety arises in describing contact interactions, so it is beneficial to work with the Lippmann–Schwinger reformulation of the Schrödinger equation and break the problem into two steps: we calculate the T-matrix (defined below) in terms of g, and then we relate the T-matrix to a; together this relates a to g.

First, we re-write the Schrödinger equation in integral form as[10]

$$|\psi\rangle = |\psi_0\rangle + \hat{G}_0 \hat{V} |\psi\rangle \tag{2.25}$$

---

[9] This depends somewhat on particle statistics and conventions; the result is given for bosons. For single particle scattering from a potential or distinguishable particles, one obtains $\sigma = 4\pi a^2$.

[10] To see this, rewrite the Schrödinger equation as $(E - H_0)|\psi\rangle = V|\psi\rangle$. Then an equation of the form $\hat{L}|\psi\rangle = |\phi\rangle$ with $\hat{L}$ a linear operator has the general solution $|\psi\rangle = |\psi_0\rangle + \hat{G}_0|\phi\rangle$ where $\hat{G}_0 \equiv \hat{L}^{-1}$ is the Green's function associated with $\hat{L}$ and $|\psi_0\rangle$ is the general solution of $\hat{L}|\psi_0\rangle = 0$. [[This is a generalization of the usual statement for differential equations, that (1) the general solution is the general solution of the homogeneous equation plus any particular solution of the inhomogeneous equation and (2) an inhomogeneous solution is given by summing over the Green's functions times the source function.]]

with $|\psi_0\rangle$ the solution for H_0, the Hamiltonian in the absence of the potential operator $\hat{V}$, and the bare Green's function defined as $\hat{G}_0 \equiv \left(E - \hat{H}_0 + i\delta\right)^{-1}$ (where the infinitesimal $\delta \to 0^+$ is chosen to give outgoing scattered waves later). We define the T−matrix via[11]

$$\hat{V}|\psi\rangle \equiv \hat{T}|\psi_0\rangle, \tag{2.26}$$

Equation 2.25 implies $V|\psi\rangle = V(1 - \hat{V}\hat{G}_0)^{-1}|\psi_0\rangle$, so the T-matrix is

$$\hat{T} = V(1 - \hat{V}\hat{G}_0)^{-1}. \tag{2.27}$$

In Fourier space this is

$$T(\mathbf{k}, \mathbf{k}', E) = V(\mathbf{k} - \mathbf{k}') + \frac{1}{\Omega}\sum_{\mathbf{k}''} V(\mathbf{k}' - \mathbf{k}'')(E - k''^2/m + i\delta)^{-1} T(\mathbf{k}, \mathbf{k}', E) \tag{2.28}$$

with Ω the volume of space. For a contact interaction, $V(\mathbf{k}) = g$, and thus Eq. 2.28 becomes

$$T(\mathbf{k}, \mathbf{k}', E) = g + \frac{g}{\Omega}\sum_{\mathbf{k}''}(E - k''^2/m + i\delta)^{-1} T(\mathbf{k}, \mathbf{k}', E). \tag{2.29}$$

One may verify that the solution to this equation is

$$T(\mathbf{k}, \mathbf{k}', E) = \frac{g}{1 - (g/\Omega)\sum_{\mathbf{k}''}(k''^2/m + i\delta)^{-1}}. \tag{2.30}$$

Appendix A shows that $a = \frac{mT(0,0,0)}{4\pi}$ so Eq. 2.30 gives

$$a = \frac{m}{4\pi}\frac{g}{1 - (g/\Omega)\sum_{\mathbf{k}''}(k''^2/m + i\delta)^{-1}}. \tag{2.31}$$

For a given physically measured a and large-momentum cutoff Λ on the **k**-sum, this tells us which g we should use in the Hamiltonian.

Note that the relationship between the scattering length a and g depends on cutoff Λ, even in the limit with $\Lambda \to \infty$. Moreover, the equation indicates $g(\Lambda)$ diverges for large Λ (for example, expanding in g, the $O(g^2)$ term diverges). This however is no problem: g is an unmeasurable quantity (independently of the cutoff Λ), and in any physical quantity, Λ and g will drop out so that only a appears in the final expressions (this is a statement of the renormalizability of the theory). As one consequence, as we will see later, if we are interested in capturing dilute limit

---

[11] There are other, equivalent, definitions of the T-matrix, but this has relevant physical content: it allows one to obtain exact quantities by using the T-matrix acting on the unperturbed wavefunction.

physics, we can do first order perturbation theory or mean field theory, use $g(\Lambda = 0) = 4\pi\hbar^2 a/m$ in place of the bare g and get the *exact* answer in this limit.

Tuning the interactions in cold atoms systems. To explore many-body physics in cold atoms, one needs to move away from the dilute limit where the kinetic energy dominates over the interaction energy. In cold atoms, there are two main experimental techniques to reach the strongly interacting regime: (1) In the first method, one boosts the interaction energy using a Feshbach resonance. With this technique, one can tune the scattering length from $a = -\infty$ to $a = +\infty$. In this technique, experimentalists use an external magnetic field to move a closed channel bound state near resonance with the open scattering channel, and the scattering length diverges at the resonance [1]. (2) In the second method, one diminishes the kinetic energy by placing the particles in an optical lattice, in which case tunneling is required to move between lattice sites, and the kinetic energy may be suppressed essentially to zero by increasing the lattice depth. Section 2.2.3 introduces this technique.

A final avenue of manipulating interactions is to use atoms with long range interactions[12] that are indescribable with a contact interaction. In experiments, this may be accomplished using ground state ultracold molecules [10], Rydberg atoms, or, on a very weak energy scale, using the magnetic dipole moment of atoms [19]. These interactions go as $1/r^3$ at long distances, and are being intensively studied right now. This is due at least partially to their ability to stabilize intermediately ordered liquids (quantum nematics) [19].

2.2.2 *Weak Interactions: Mean Field Theory and Fluctuations*

Section 2.1.1 discussed the Bose–Einstein condensation of the non-interacting Bose gas. Here I discuss how weak interactions alter this picture. In particular, I will consider the effect of the interaction on the ground state energy and on the ground state's structure. The Hamiltonian describing the dilute Bose gas is

---

[12] The definition of "long ranged" is dependent on context: whether a given potential counts as long ranged depends on the physics one is interested in. For example, specifying to three dimensions, there are at least two senses in which a $1/r^3$ potential is long ranged: (1) in the thermodynamic limit, the interaction energy grows faster than extensively for a uniform system and (2) it is indescribable by a contact interaction at low energies (note it is the *fastest* decaying interaction indescribable by a contact interaction at low energies, since all $1/r^\alpha$ with $\alpha > 3$ may be described as contact interactions at low energies). However, it is *not* long ranged in the sense that it decays faster than the $1/(\text{length})^2$ characteristic of kinetic energy. Consequently, the high (low) density limit is still strongly (weakly) interacting, in contrast to say the Coulombic interaction, where this is switched (e.g., one forms a Wigner crystal at low density and a Fermi liquid at high density). A final subtlety is that although some other interactions decay slower, for example the Coulomb interaction, they can still have extensive interaction energy in the thermodynamic limit because they are compensated systems: positive and negative charges screen each other at long distance.

$$H = \int d\mathbf{r}\left[-\psi^\dagger(\mathbf{r})\frac{\nabla^2}{2m}\psi^\dagger(\mathbf{r})\right] + \frac{g}{2}\int d\mathbf{r}\psi^\dagger(\mathbf{r})\psi^\dagger(\mathbf{r})\psi(\mathbf{r})\psi(\mathbf{r}), \tag{2.32}$$

with an implicit momentum cutoff and the contact interaction g appropriately renormalized, as discussed in Sect. 2.2.1. In Fourier space, this is

$$H = \sum_{\mathbf{k}} \frac{k^2}{2m} b^\dagger_{\mathbf{k}} b_{\mathbf{k}} + \frac{g}{2V}\sum_{\mathbf{k},\mathbf{k}',\mathbf{q}} b^\dagger_{\mathbf{k}+\mathbf{q}} b^\dagger_{\mathbf{k}'-\mathbf{q}} b_{\mathbf{k}'} b_{\mathbf{k}}. \tag{2.33}$$

Since we are interested in weak interactions, we perturb in g. We know that for $g = 0$ the ground state $|N\rangle$ is that of N particles in the $k = 0$ state: $|N\rangle = \mathcal{N}(\psi^\dagger_{\mathbf{k}=0})^N|\text{vac}\rangle$ for some normalization factor $\mathcal{N}$ with $|\text{vac}\rangle$ the vacuum state. The first order correction to the ground state energy is then read off from Eq. 2.33 to be

$$\langle N|H|N\rangle = \frac{gN(N-1)}{2V} \tag{2.34}$$

In the thermodynamic limit, this is

$$\langle E\rangle = N\frac{gn}{2}, \tag{2.35}$$

with $n \equiv N/V$, which gives the energy shift due to weak interactions.

Section 2.2.1 showed that g was a function of the cutoff used. However, we never specified a cutoff, and our naive perturbative calculation yields an energy $g(\Lambda)n/2$ as the energy density, independent of Λ. Since g depends on Λ, this looks like a problem: our answer depends on the cutoff. Deciding on the proper Λ requires a little work.

First, I give an intuitive picture of why we should take $g(\Lambda = 0)$, then describe a more formal route (for more discussion, see Refs. [1, 17]). Intuitively, we want to use zero cutoff, so that already at the first order level, we have included the effects of all two-body scattering. More formally, we keep a finite cutoff in the calculation, and expand perturbatively in g. The cutoff dependence drops out and we are led to the same conclusion: we should use the g corresponding to zero cutoff.[13]

This perturbative result is also obtained by treating the interaction with mean field theory, and assuming that the system macroscopically occupies some mode. Such a picture also allows one to treat the spatially varying case, in which case one obtains a nonlinear eigenvalue equation—the *Gross-Pitaevskii equation*,

$$E\psi(\mathbf{r}) = \psi^*(\mathbf{r})\left(-\frac{1}{2m}\nabla^2 + V(\mathbf{r})\right)\psi(\mathbf{r}) + g|\psi(\mathbf{r})|^2\psi(\mathbf{r}), \tag{2.36}$$

---

[13] If we tried the same trick expanding perturbatively in $g(\Lambda = \infty)$, the second order correction would diverge.

where $\psi(\mathbf{r})$ is the condensate wavefunction and $V(\mathbf{r})$ is a possible external potential.

In addition to the energy, it is also instructive to consider the interactions' effect on the ground state itself. One consequence is a depletion of the condensate. At zero temperature for the non-interacting system all of the particles were in the ground state single-particle wavefunction, the $\mathbf{k} = 0$ wavefunction for homogeneous systems. Interactions cause some fraction of these atoms to scatter into higher momenta states. The lowest order depletion is calculated by replacing the zero momentum field operator with a c-number: $b_{\mathbf{k}=0} \approx N_0$ for some number N_0 (not necessarily the total particle number), and expanding the Hamiltonian to second order in the $b_{\mathbf{k}}$ for $\mathbf{k} \neq 0$. In this "Boguliobov approximation," one finds [1]

$$\frac{N - N_0}{N} = \frac{8}{3\sqrt{\pi}}\left(na^3\right)^{1/2} \tag{2.37}$$

and we see that the condensate fraction, $\langle b^\dagger_{\mathbf{k}=0} b_{\mathbf{k}=0}\rangle / \sum_{\mathbf{k}} \langle b^\dagger_{\mathbf{k}} b_{\mathbf{k}}\rangle = N_0/N$, is less than unity.

Thus, increasing interaction strength depletes the condensate. It is natural to ask if it is possible to completely deplete the condensate, and if so what happens. The following section discusses this question for interactions in a deep lattice.

2.2.3 *Bosons in an Optical Lattice and the Bose–Hubbard Model*

I will argue that bosonic atoms in the sinusoidal optical lattice potential given by Equation 2.16 are frequently well-described by the Bose–Hubbard model [20], whose Hamiltonian is

$$H = -t\sum_{\langle i,j\rangle} b_i^\dagger b_j + \frac{U}{2}\sum_i b_i^\dagger b_i^\dagger b_i b_i \tag{2.38}$$

where the $\sum_{\langle i,j\rangle}$ runs over nearest neighbor pairs i,j, and b_i and $b_i^\dagger$ are bosonic annihilation and creation operators, satisfying $[b_i, b_j] = 0$ and $[b_i, b_j^\dagger] = \delta_{ij}$. The Bose–Hubbard model is a canonical model of solid state physics introduced in the context of granular superconductors. I will also give the conditions under which the Bose–Hubbard model accurately describes cold atomic systems. This argument was first given by Jaksch et al. [21]. Although the Bose–Hubbard model is arguably the simplest lattice model showing phase transitions,[14] the physics is quite rich and there are even quite important open questions in the quantum critical

[14] Beyond the fairly trivial transitions from finite density to the vacuum that occur even in non-interacting continuum systems.

regime (see Chap. 13). A basic introduction is given in Sect. 2.2.4 and more is discussed throughout the thesis.

The same concept—that ultracold atoms in optical lattices emulate important models of solid state physics—is being used to realize the Fermi–Hubbard model,[15] a central model in strongly correlated electron systems and which contains much of the physics of the high temperature superconducting compounds, in cold atoms. The community is exploring many other possibilities, as discussed in the introduction, and the prospects are exciting.

A simple physical picture to understand the applicability of the Bose–Hubbard model comes from thinking about deep lattices, where sites decouple. In this limit, each site may be solved individually, and one obtains a tightly localized spatial wavefunction at each site. In the absence of interactions, we obtain a series of discrete energy levels, roughly harmonic oscillator like. For weak interactions, adding particles to the site just multiply occupies the lowest energy on-site eigenstate, and this determines an on-site interaction energy. The most important perturbation from this limit is the ability for particles to tunnel to neighboring sites, and this is captured by the tunneling term in the Hamiltonian.

To capture this idea more precisely and formally, one constructs a set of Wannier functions [22]. One can define the Wannier functions $w(\mathbf{r})$ by

$$w_\alpha(\mathbf{r}) \equiv \frac{1}{\sqrt{N}} \sum_{\mathbf{k}\in \mathrm{BZ}} \phi_{\alpha,\mathbf{k}}(\mathbf{r}) \tag{2.39}$$

where $\phi_{\alpha,\mathbf{k}}$ is the quasi-momentum $\mathbf{k}$ Bloch function[16] for the αth bandand the sum runs over first Brillouin zone, N is the total number of lattice sites, and the sum runs over first Brillouin zone. Using the properties of the Bloch functions, one can show that they may be written in terms of the Wannier function as

$$\phi_{\alpha,\mathbf{k}}(\mathbf{r}) = \frac{1}{\sqrt{N}} \sum_{\mathbf{R}} e^{i\mathbf{k}\cdot\mathbf{R}} w_\alpha(\mathbf{r} - \mathbf{R}) \tag{2.40}$$

where the sum runs over all lattice vectors $\mathbf{R}$. Thus we see that the Wannier states $w_\alpha(\mathbf{r} - \mathbf{R})$ for all lattice vectors $\mathbf{R}$ form a complete basis for the αth band. Consequently, we can write our lattice Hamiltonian in the basis of Wannier functions and we are justified in describing the system using only the lowest band Wannier functions when the band gap to the first excited band is sufficiently large that few excitations to this band are relevant. We will examine the condition under which this applies below. We associate the creation and annihilation operators with the Wannier states as

[15] Just called "The Hubbard model" in virtually all other areas of physics.

[16] Bloch functions are just the eigenstates of a non-interacting periodic system. Symmetry dictates that they are periodic under translations by lattice vectors $\mathbf{R}$ up to a phase, from which the rest of the properties follow.

$$c_j = \int d\mathbf{r}\, w(\mathbf{r} - \mathbf{R_j})\psi(\mathbf{r}) \tag{2.41}$$

with $\psi(\mathbf{r})$ the annihilation operator at a position $\mathbf{r}$. Because the Wannier functions turn out to be short ranged under quite general conditions—certainly for deep lattices—and this justifies keeping only the short ranged terms in the resulting Hamiltonian for the lowest band, one obtains the Bose–Hubbard Hamiltonian with parameters [21]

$$t = \int d\mathbf{r} w^*(\mathbf{r}) \left[\frac{-\hbar^2}{2m} \nabla^2 + \sum_{\alpha=x,y,z} V_\alpha \sin(\mathbf{k}_\alpha \cdot \mathbf{r}) \right] w(\mathbf{r})$$

$$U = \frac{4\pi\hbar^2 a}{m} \int d\mathbf{r} |w(\mathbf{r})|^4. \tag{2.42}$$

The first requirements for the Bose–Hubbard description to be valid—that the band gap is large compared to interactions (and temperature)—requires that the interaction energy density, $\propto a\rho^1/3$, with ρ the maximum three dimensional density, is small compared to the kinetic energy density $\propto \rho^{2/3}$, or $a\rho^{-1/3} \ll 1$. For deep lattices, particles are put in the lowest harmonic oscillator state associated with oscillations of frequency Ω around the potential minimum at each site, and the density is thus n/ℓ^3 where $\ell = \sqrt{\hbar/(m\Omega)}$ is the harmonic oscillator length and n is the site occupation. Then the requirement that the interaction effects are describable with a single band is that $a \ll \ell$. The criterion that Wannier functions are short range requires that $\ell \ll d$, where d is the lattice spacing. So, to summarize, the Bose–Hubbard description is valid for

$$a \ll \ell \ll d. \tag{2.43}$$

For typical numbers, $a = 5$ nm for ^{87}Rb and $d = 500$ nm, so that we can find lattice depths satisfying these criteria.

2.2.4 Gutzwiller Mean Field

In this section, I will discuss the qualitative physics of the Bose–Hubbard model, Eq. 2.38. Here and in Sect. 2.2.5 I will also introduce some theoretical approaches to describing the many-body physics of this system and discuss how the required formalism is somewhat different than that introduced in typical many-body (e.g., diagrammatic) formalisms.

To obtain a global picture of the system's behavior, I will first consider two limits: very weak and very deep lattices. Note that there are only two energy scales in the Hamiltonian, t and U, and thus up to an overall scale, the physics depends only on the ratio t/U. The weak and deep lattice limits correspond to $t/U = \infty$ and $t/U = 0$ respectively. In the weak limit, interactions play no role. As long as the temperature is much less than the bandwidth ($\sim t$) the system occupies only low momentum states, and the dispersion may be approximated as quadratic. Then the physics reduces to the superfluidity of nearly free particles. In the deep lattice limit, the absence of tunneling implies that can treat sites independently, and the eigenstates are Fock states with fixed particle occupation n at each site. For small t/U, if the average occupation is an integer, then there is a large energy cost ($\sim U$) to moving a particle from one lattice site to another, which the kinetic energy cannot overcome: the system is a Mott insulator. In between there is a phase transition. The phase diagram and physics of the phases at zero and finite temperature are a recurring theme explored throughout this thesis.

A typical approach to the many-body problem is to try to use perturbation theory from either the weak or strong coupling limits, where the behavior is known. However, perturbing from one limit is guaranteed to fail to describe the other phase because the perturbation theory guarantees analytic behavior in the expansion parameter, while the phase transition is signified by non-analyticities of observables.[17] However, there is a simple mean field theory capable of qualitatively capturing both limits. This "Gutzwiller mean field theory" [20, 21] is described in this section. Although it is simple, its formal structure is quite different than traditional mean field theories, such as Hartree–Fock, BCS, etc., and going beyond the mean field theory is non-trivial, discussed in Sect. 2.2.5. Although sophisticated numerical algorithms such as the worm algorithm [23] can obtain high accuracy numerical results for static behavior, no method exists to quantitatively calculate dynamics across the full phase diagram (especially in the quantum critical regime).

There are several ways to formulate the Gutzwiller mean field method given here, as with other mean field methods. One may approximate some operator in the Hamiltonian by its average value (mean field approach), construct a variational wavefunction, resum some infinite set of diagrams, truncate operator equations of motion, or make some approximation to the path integral, just as some examples. Here, I will give the variational formulation, and then briefly interpret it as a mean field. Chapters 13 and 14 introduce and make use of the path integral formulation, while Chap. 5 relies on an equation of motion technique to capture dynamics.

The Gutzwiller variational ansatz takes each site to be uncorrelated with the other sites. That is, the variational wavefunction $|\Psi\rangle$ is a tensor product over all lattice sites of wavefunctions on each site:

[17] An exciting alternative perspective to understand both regimes is to do perturbation theory around the phase transition point, and this (much more difficult) perturbation theory is essentially the basis of describing quantum critical systems.

$$|\Psi\rangle = \bigotimes_i \left(\sum_{n=0}^{\infty} f_{n,i} |n\rangle_i \right) \tag{2.44}$$

where $|n\rangle_i$ is the number occupation n eigenstate for site i. We assume that translational invariance is not spontaneously broken, so that $f_{n,i}$ is independent of the site index i. Minimizing the Bose–Hubbard model, Eq. 2.38's, energy expectation value in this state gives equations for the f_n's:

$$\epsilon f_n = -tz\left(\sqrt{n}\alpha^* f_{n-1} + \sqrt{n+1}\alpha\, f_{n+1}\right) + \left(\frac{U}{2}n(n-1) - \mu n\right) f_n \tag{2.45}$$

with ϵ the eigenvalue of this problem, z the lattice coordination number, and

$$\alpha = \sum_n \sqrt{n} f_n^* f_{n-1}. \tag{2.46}$$

These are coupled non-linear algebraic equations for the f_n's. To solve them, we take a guess for α, and solve the first equation for the f_n's by truncating to a reasonable number of occupations. Next, we determine a new α from the second equation. We then repeat this procedure until it converges. Even deep in the superfluid, the number of states kept need only be $\sim \sqrt{\langle n \rangle}$ where, and for the low filling states usually of interest, with $\langle n \rangle \lesssim 3$, truncating to occupations $n = 0, 1, 2, 3, 4, 5$ is generally sufficient to get many decimal places of accuracy. Even keeping much large range of occupations, this can be done virtually instantaneously with any modern computer. In the limit where one has small t/U, one can generically truncate to two occupations, $|n\rangle$ and $|n+1\rangle$, and solve the equations analytically. Note that these equations are identical to those obtained for the eigenvalue problem obtained by approximating the Hamiltonian Eq. 2.38 by writing $\sum_{\langle i,j \rangle} a_i^\dagger a_j \approx z\langle a_j \rangle \sum_i a_i^\dagger$.

A key feature of the solutions to these equations is that they capture the Mott insulator to superfluid phase transition. The system displays a non-analytic behavior, and thus a phase transition: below a critical t/U for any given μ, the solutions are just the number eigenstates: $f_n = \delta_{nm}$ for some m. At a critical t/U, small number fluctuations start upon increasing t/U (entering the superfluid) and get larger as one goes deeper into the superfluid (larger t/U).

The Gutzwiller approximation is exact when $t = 0$—it can represent number eigenstates. It is also exact in infinite dimension,[18] since in this case one site feels the influence of infinitely many neighbors, and any fluctuations of the mean field average out. Remarkably, it is also exact deep in the superfluid (in the

[18] Although this holds generally in classical mechanics, the issue in quantum mechanics is somewhat trickier. Thinking of a path integral description, one needs to worry about the time dimension. The proper description of the infinite spatial dimension (but still one real or imaginary time dimension) requires *dynamical mean field theory*, which for bosonic systems is presented in Ref. [24].

non-interacting limit): if one takes the f_n's to yield coherent states at each site, and project the resulting Gutzwiller wavefunction onto a state of fixed *total* particle number N, one obtains exactly the non-interacting ground state, with N particles in the $k = 0$ single particle ground state. The fact that the ansatz exactly captures both limits' behavior is remarkable, but it is worth noting that it captures physics away from these limits only qualitatively—even for arbitrarily small deviations from the limits. Numerically, it is has been confirmed to give qualitatively quite accurate phase diagrams in $d = 1, 2, 3$. For example, the unity site filling Mott insulator/superfluid transition on a three-dimensional cubic lattice occurs at $(t/U)_c = 0.03408(2)$ while GMFT yields $(t/U)_c = 0.029$ [25].

2.2.5 *Fluctuations in Strong Lattice Potentials: Beyond Gutzwiller*

The Gutzwiller approximation captures on-site number fluctuations and infinite range condensate order, but fails to capture intermediate range correlations. Thus, although it is exact in the deep Mott limit and superfluid limit, it fails to capture even infinitesimal deviations from these limits exactly. For example, in the dilute superfluid limit the Gutzwiller approximation captures the condensate depletion only approximately, failing to quantitatively reproduce the Boguliubov result above.

There are several approaches to incorporating intermediate range fluctuations around the Gutzwiller ansatz. One is to abandon Gutzwiller in favor of a strong coupling perturbation theory in t/U. Remarkably, when resummed with a Pade determinant—approximating the observables as

$$f(\tilde{t}) = \frac{\sum_{i=0}^{i_{\max}} \tilde{t}^i}{\sum_{j=0}^{j_{\max}} \tilde{t}^j}, \tag{2.47}$$

with the definition $\tilde{t} \equiv t/U$, and truncating sums in both the numerator ($i_{\max}$) and denominator ($j_{\max}$) beyond lowest order. Then low order perturbation theory can reproduce numerically exact quantum Monte Carlo simulations for systems not too deep in the superfluid [26, 27].

A more global approach is to include fluctuations around the Gutzwiller approximation. This is an appealing approach because it can extend the strength of the Gutzwiller approach—its exactness in both the small and large t/U and high dimensions—to additionally capture the intermediate range correlations and momentum space structure of the dispersion. However, it is a technically difficult calculation. The broad idea is that the Gutzwiller ansatz forms the vacuum around which one wants to include fluctuations, and this vacuum's structure is non-trivial. Specifically, fields operators in this state do *not* satisfy Wick's theorem; in contrast to equilibrium states of quadratic Hamiltonians, the more complicated structure of the Gutzwiller ansatz prevents Wick's theorem from applying.

Although the ansatz provides some simplifications of its own—for example, expectations of products of operators factor if the operators are on different sites—this non-triviality means that the standard techniques of many body theory fail, since these all rely on the applicability of Wick's theorem to the vacuum.[19] For similar reasons, even linear response dynamics are non-trivial, and Chap. 5 gives an approximate solution to one linear response problem—radio-frequency spectroscopy.

2.2.6 Extracting Information About Homogeneous Systems from Trapped Systems

In condensed matter physics we usually wish to understand macroscopically homogeneous samples. Cold atomic system are invariably trapped, and are at most hundreds of lattice sites across in this spatially varying potential. Here I discuss the conditions under which it is possible to extract information about a homogeneous system from a cold atoms system, and how one may do this.

In the limit where the potential varies infinitely slowly, it is clear that we can break the system up into macroscopic regions, in each of which the potential varies sufficiently little to be treated as homogeneous. Thus, measurements in each region give properties of the appropriate homogeneous system. We can absorb the spatially varying potential $V(\mathbf{r})$ into the free energy to define a spatially varying chemical potential $\mu(\mathbf{r}) \equiv \mu_c - V(\mathbf{r})$. Then in the limit where the trap varies sufficiently slowly, the system at point $\mathbf{r}$ shares all its local properties with those of a homogeneous with chemical potential $\mu(\mathbf{r})$. This approximation is called the Thomas–Fermi or local density approximation (LDA),[20] and was originally introduced to approximate the properties of atoms by treating the electrons around nuclei in this manner.

Define

$$L \equiv \mu(r)/\mu'(r), \tag{2.48}$$

[19] Even symmetry breaking causes some difficulties, since the vacuum is altered, but these can generally be alleviated by perturbing around the quadratic theory applicable near the new symmetry broken solution.

[20] Not to be confused with the much more sophisticated local density approximation used in the electronic structure community. Only the *corrections* to the non-interacting quantum mechanics problem are treated with a local approximation there.

which is a measure of the characteristic length scale on which the chemical potential varies by an amount comparable to its value.[21] If the effective chemical potential varies negligibly over a correlation length, then different regions with this small chemical potential difference are independent when separated by scales much more than that. That is, the Thomas–Fermi approximation is valid if

$$\xi \ll L \tag{2.49}$$

where ξ is the longest correlation length of the system. In typical systems, $L \sim 10$–20, and can be made larger in some regions by engineering the shape of $V(r)$. Deep in phases of matter (far from transitions), ξ is typically a few lattice spacings. When ξ is much larger than the lattice spacing, as happens near critical points, universal behavior manifests. This phenomena has the largest correlation lengths of interest,[22] and since one can explore ξ up to $\sim L \gg 1$, there is even a window into even this regime. To summarize, we expect that we can obtain rather accurate information of homogeneous system over whole phase diagram, even where the physics has reached its long wavelength critical behavior.

From another vantage point, we can ask how obtaining information about systems that are only homogeneous on scales $\sim L$ compares with numerically calculating the properties on a computer. Although L may seem small, recall that even for the simplest systems, e.g., simple spin-1/2 lattice models, our arguments in the introduction indicated that calculating general properties of a $10 \times 10 \times 10$ system, even with sophisticated numerical algorithms, supercomputers, and years of patience, is impossible. Thus, measurements of trapped systems can give otherwise intractable information about a homogeneous system's behaviors.

There is (at least) one important caveat to this conclusion, however. In systems with spontaneously formed inhomogeneities (such as charge density waves or vortex lattices), there is an ordering length scale which may be much larger than the lattice spacing. Even if this order is fluctuating—that is, has no static infinitely long ranged correlations—as may happen when multiple phases compete, one needs to worry about this. In this case, the Thomas–Fermi approximation is inapplicable under typical trapping conditions.

References

1. C.J. Pethick, H. Smith, *Bose–Einstein Condensation in Dilute Gases* (Cambridge University Press, Cambridge, 2001)
2. H.J. Metcalf, P. vander Straten, P. Straten, *Laser Cooling and Trapping*, corrected edn. (Springer, London, 1999)

[21] For many cases of interest, the value of the chemical potential is also on the order of all the interesting scales in the problem, e.g. the tunneling rate t and interaction energy U. Other circumstances may be treated analogously.

[22] Once one understands the physics of this regime, one can extrapolate to infinitely large ξ to obtain a complete picture of behavior throughout a system's phase diagram.

3. F.P.D. Santos, J. Lonard, J. Wang, C.J. Barrelet, F. Perales, E. Rasel, C.S. Unnikrishnan, M. Leduc, C. Cohen-Tannoudji, Phys. Rev. Lett. **86**, 3459 (2001)
4. A. Robert, O. Sirjean, A. Browaeys, J. Poupard, S. Nowak, D. Boiron, C.I. Westbrook, A. Aspect, Science **292**, 461 (2001)
5. A. Griesmaier, J. Werner, S. Hensler, J. Stuhler, T. Pfau, Phys. Rev. Lett. **94**, 160401 (2005)
6. Y. Takasu, K. Maki, K. Komori, T. Takano, K. Honda, M. Kumakura , T. Yabuzaki, Y. Takahashi, Phys. Rev. Lett. **91**, 040404 (2003)
7. S. Kraft, F. Vogt, O. Appel, F. Riehle , U. Sterr, Phys. Rev. Lett. **103**, 130401 (2009)
8. S. Stellmer, M.K. Tey, B. Huang, R. Grimm , F. Schreck, Phys. Rev. Lett. **103**, 200401 (2009)
9. Y.N.M. de Escobar, P.G. Mickelson, M. Yan, B.J. DeSalvo, S.B. Nagel, T.C. Killian, Phys. Rev. Lett. **103**, 200402 (2009)
10. K. Ni, S. Ospelkaus, M.H.G. de Miranda, A. Pe'er, B. Neyenhuis, J.J. Zirbel, S. Kotochigova, P.S. Julienne, D.S. Jin, J. Ye, Science **322**, 231 (2008)
11. A. Ashkin, Proc. Natl Acad. Sci. USA **94**, 4853 (1997)
12. M. Ueda, Y. Kawaguchi, arxiv:1001.2072 (2010) (preprint)
13. W.S. Bakr, A. Peng, M.E. Tai, R. Ma, J. Simon, J.I. Gillen, S. Flling, L. Pollet, M. Greiner, Science **329**, 547 (2010)
14. K.D. Nelson, X. Li, D.S. Weiss, Nat. Phys. **3**, 556 (2007)
15. T. Gericke, P. Wurtz, D. Reitz, T. Langen, H. Ott, Nat. Phys. **4**, 949 (2008)
16. N. Gemelke, X. Zhang, C. Hung, C. Chin, Nature **460**, 995 (2009)
17. A.A. Abrikosov, *Methods of Quantum Field Theory in Statistical Physics*, revised edn. (Dover Publications, New York, 1975)
18. L.D. Landau, L.M. Lifshitz, *Quantum Mechanics Non-Relativistic Theory*, 3rd edn, vol. 3. (Butterworth-Heinemann, London, 1981)
19. T. Lahaye, C. Menotti, L. Santos, M. Lewenstein, T. Pfau, Rep. Prog. Phys. **72**, 126401 (2009)
20. M.P.A. Fisher, P.B. Weichman, G. Grinstein, D.S. Fisher, Phys. Rev. B **40**, 546570 (1989)
21. D. Jaksch, C. Bruder, J.I. Cirac, C.W. Gardiner, P. Zoller, Phys. Rev. Lett. **81**, 3108 (1998)
22. N.W. Ashcroft, N.D. Mermin, *Solid State Physics*, 1st edn. ((Brooks Cole, Belmont, 1976)
23. N. Prokofev, B. Svistunov, I. Tupitsyn, J. Exp. Theor. Phys. **87**, 310 (1998)
24. K. Byczuk, D. Vollhardt, Phys. Rev. B **77**, 235106 (2008)
25. B. Capogrosso-Sansone, N.V. Prokof'ev, B.V. Svistunov, Phys. Rev. B **75**, 134302 (2007)
26. J.K. Freericks, H. Monien, Phys. Rev. B **53**, 2691 (1996)
27. J.K. Freericks, H.R. Krishnamurthy, Y. Kato, N. Kawashima, N. Trivedi, Phys. Rev. A **79**, 053631 (2009)

Chapter 3
Radio-Frequency Spectroscopy: Broad Introduction

3.1 Motivation and Background

Radio-frequency (RF) spectroscopy has emerged as a powerful probe of cold atoms. It provided the first evidence of BEC in dilute hydrogen gases [1], probed the pair structure and binding energy across the BEC/BCS crossover [2], and was used to study the quantum phase transition between Mott and superfluid states of lattice boson [3].

Generally speaking, spectra are useful because they probe excitation structure, which is hidden in density profiles and time-of-flight measurements common in cold atoms. That is, they tell what happens when particles and energy are added to or removed from a system in various manners. This is moreover connected to the dynamic correlations present in the initial state.

In cold atoms, RF spectra are particularly feasible, compared to other spectroscopic probes, since in a typical cold atoms setup, the only extra necessary equipment is an RF coil, and this often is already present, for evaporative cooling or other purposes. In RF spectroscopy, an RF photon is absorbed to alter the internal state of an atom. Usually, this is a transition between the atom's hyperfine energy levels. In practice, experimentalists may utilize other transitions such as microwave or optical transitions, but I will refer to these as "RF spectra" as long as the recoil momentum is negligible.[1]

As argued above, RF spectra are relatively experimentally feasible and provide significant information about correlations and excitations. Also, because of their low momentum transfer, they can be used to probe spectral lines in a momentum-insensitive, "Doppler free" manner. However, the flip side of this is that measuring momentum dependence of excitations is impossible. For this

[1] This can be accomplished by using two counter-propagating lasers to drive an optical transition, for example. Because this eliminates momentum recoil, there is no Doppler broadening of the line and is referred to as Doppler-free spectroscopy

K. R. A. Hazzard, *Quantum Phase Transitions in Cold Atoms and Low Temperature Solids*, Springer Theses, DOI: 10.1007/978-1-4419-8179-0_3,

reason, the probe is less powerful than Bragg or Raman spectroscopy, although these are somewhat more difficult experimentally. It is worth pointing out that if one can do momentum resolved Bragg spectra, then one can more or less just as easily do Raman spectroscopy, for which RF spectroscopy is the zero-momentum limit. Thus understanding RF spectra also offer a first step to understanding these spectra. An alternative method to obtain momentum resolution using RF spectra is to do time-of-flight imaging after applying the RF pulse. Debbie Jin's group has demonstrated this on strongly interacting Fermi gases in the BEC/BCS crossover [4], but this method allows one to measure only trap averaged spectra.

3.2 Two Differing Pictures of RF Spectroscopy

RF spectra probe cold atoms' correlations and excitation structure. However, two pictures, both commonly used in the literature, suggest very different results for how these correlations and excitations manifest themselves. The first picture is based on sum rules, while the second is based on the single particle Green's function. In the next few chapters I reconcile these as limiting cases of a more complete theory.

Sum rule picture. In a standard application of sum rules to understand spectra, one derives an expression for the average spectral line shift $\langle\delta\omega\rangle$ due to interactions with other atoms, and then assumes the spectrum is a single δ-function line with the associated spectral location. This gives an accurate assessment of many situations such as dilute gases [5] and quantifies the density inhomogeneity-limited accuracy of atomic clocks [6, 7]. Chapter 4 fleshes out this picture in relation to experiments at MIT for lattice bosons near the Mott insulator/superfluid quantum phase transition.

Single particle Green's functions. Another common approach to calculate RF spectra is to relate them to single particle spectra. In the limit where there are no final state interactions—that is, the final internal state of the atom does not interact with atoms in the initial internal state—the RF spectrum is a convolution of the zero-momentum part of the hole single particle Green's function (and in special cases is simply the momentum-averaged single particle Green's function). In contrast to the sum rule approach, this leads to quite a different picture for the superfluid's RF spectra: inserting a hole into the system can excite gapless phonons, but can also excite gapped amplitude modes.

Remarkably, Chap. 5 demonstrates that each of these pictures emerges in separate limits of a more complete calculation. I give a simple physical explanation of this, and believe that the basic point will persist to other many-body systems.

After this, Chap. 6 considers the finite temperature spectra in the deep Mott and superfluid phases. The finite temperature quantum critical region is discussed later, in Chap. 13.

References

1. D.G. Fried, T.C. Killian, L. Willmann, D. Landhuis, S.C. Moss, D. Kleppner, T.J. Greytak, Phys. Rev. Lett. **81**, 3811 (1998)
2. I. Bloch, Nat. Phys. **1**, 23 (2005)
3. G.K. Campbell, J. Mun, M. Boyd, P. Medley, A.E. Leanhardt, L.G. Marcassa, D.E. Pritchard, W. Ketterle, Science **313**, 649 (2006)
4. G. Veeravalli, E. Kuhnle, P. Dyke, C.J. Vale, Phys. Rev. Lett. **101**, 250403 (2008)
5. C.J. Pethick, H.T.C. Stoof, Phys. Rev. A **64**, 013618 (2001)
6. K. Gibble, S. Chu, Phys. Rev. Lett. **70**, 1771 (1993)
7. C. Fertig, K. Gibble, Phys. Rev. Lett. **85**, 1622 (2000)

Chapter 4
RF Spectra: A Sum Rule Approach to Trapped Bosons in an Optical Lattice

4.1 Chapter Overview

We calculate the interaction induced inhomogeneous broadening of spectral lines in a trapped Bose gas as a function of the depth of a three-dimensional cubic optical lattice. As observed in recent experiments, we find that the terraced "wedding-cake" structure of Mott plateaus splits the spectrum into a series of discrete peaks. The spectra are extremely sensitive to density corrugations and trap anharmonicities. For example, even when the majority of the cloud is superfluid the spectrum displays discrete peaks.

4.2 Introduction

The study of quantum degenerate atoms confined to periodic potentials forms an important subfield of modern atomic physics. Research in this area is driven by its connection to condensed matter physics and quantum information processing [1, 2]. A rich set of probes, including optical spectroscopy, noise spectroscopy, interference, and density profile measurements [3–8], have been used to characterize these systems, with a focus on understanding the interaction driven superfluid–insulator transition. Here we analyze in detail what information one gains from inhomogeneous pressure shifts of spectral lines in a gas of bosons confined to an optical lattice.

Atomic interactions lead to pressure and density dependent shifts of atomic lines. These "clock shifts" limit the accuracy of atomic clocks. In an inhomogeneous

This chapter was adapted from "Hyperfine spectra of trapped bosons in optical lattices" by Kaden. R. A. Hazzard and Erich. J. Mueller, published in Physical Review A **76**, 063612 (2007).

K. R. A. Hazzard, *Quantum Phase Transitions in Cold Atoms and Low Temperature Solids*, Springer Theses, DOI: 10.1007/978-1-4419-8179-0_4,

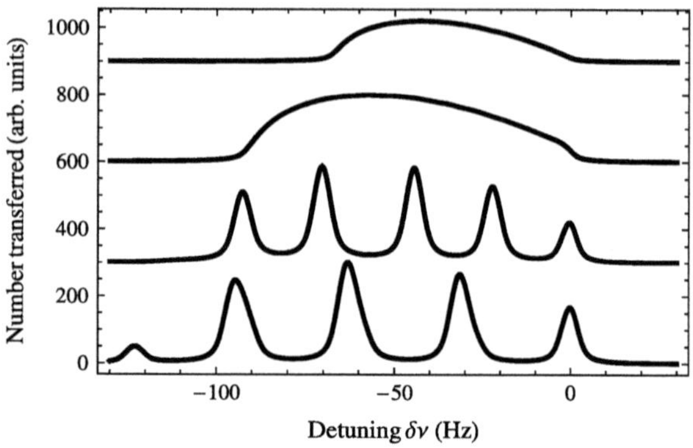

Fig. 4.1 Theoretical spectra showing the number of ^{87}Rb atoms transferred from hyperfine state $|a\rangle = |F = 1, m_F = -1\rangle$ to state $|b\rangle = |F = 2, m_F = 1\rangle$ when excited by light detuned from resonance by the frequency $\delta\nu$. The $N = 9 \times 10^4$ atoms are confined by a harmonic potential with $\bar{\omega} = (\omega_x\omega_y\omega_z)^{1/3} = 2\pi \times 100$ Hz and a three-dimensional periodic potential with lattice depth $V_0 = 5$, 10, 25, $35E_{\text{rec}}$ (from *top* to *bottom*)

system they are spatially dependent, yielding a broadened spectrum whose structure reveals details about the local atomic correlations. In many situations the clock shift is proportional to the atomic density, and the spectral line directly gives a histogram of the atomic density. As an example of this technique, Bose–Einstein condensation in spin polarized atomic hydrogen was detected through the line shape of a two-photon 1s–2s transition [9]. More recently, Campbell et al. [8] utilized atomic clock shifts to experimentally probe bosons trapped in an optical lattice, finding evidence for Mott insulating shells. Motivated by these latter experiments, we present a theoretical analysis of the lineshapes which should be found when bosonic atoms in a periodic potential are confined by a nominally harmonic potential.

In Sect. 4.3 we use a local density approximation to calculate the spectrum of a harmonically trapped gas as a function of the depth of an optical lattice (Fig. 4.1). In the deep lattice limit, the spectral line splits into several distinct peaks, associated with the formation of density plateaus. Due to the sensitivity of these spectra to small density corrugations, this splitting occurs even when large sections of the cloud are superfluid. Despite qualitative agreement with experiments, our calculation severely underestimates the small detuning spectral weight. In Sect. 4.4 we show that these discrepancies are consistent with trap anharmonicities. We also explore other mechanisms for enhancing the low detuning spectral weight.

Experimental details. Since we are largely concerned with the experiment in Ref. [8], we give a brief review of the important experimental details. In these experiments, a gas of ^{87}Rb atoms in the $|a\rangle = |F = 1, m_F = -1\rangle$ hyperfine state (F the total spin and m_F its z component) was cooled well below the condensation temperature (W. Ketterle, 2007, private communication). By combining optical and magnetic fields, a three-dimensional periodic potential $V_{\text{per}} = -V_0[\cos(2\pi x/d) + \cos(2\pi y/d) + \cos(2\pi z/d)]$ was superimposed on a trapping potential. The spacing between lattice sites, $d = \lambda/2 = 532$ nm, is half of the lattice lasers' wavelength. The lattice depth V_0 was tuned from zero to $40E_{\text{rec}}$where $E_{\text{rec}} = \frac{\hbar^2}{2m}\left(\frac{2\pi}{\lambda}\right)^2$ is the photon recoil energy. A microwave and RF field were tuned near resonance for a two photon transition from the $|a\rangle$ state to an excited hyperfine state $|b\rangle = |F = 2, m_F = 1\rangle$ [10]. The number of atoms transferred during a fixed time interval was studied as a function of the RF frequency.

4.3 Spectrum of Harmonically Trapped Gas

4.3.1 Hamiltonian and Approximations

Hamiltonian. Bosons in a sufficiently deep optical lattice are described by the Bose–Hubbard model [11], found by projecting the full Hamiltonian onto the lowest Bloch band. We will work with a two-internal state Bose–Hubbard Hamiltonian, where a_i and b_i annihilate bosons at site i in states $|a\rangle$ and $|b\rangle$, respectively. Including an external trapping potential the Hamiltonian is

$$\begin{aligned} H = &- t_a \sum_{\langle i,j \rangle} a_i^\dagger a_j + \sum_i \left[\frac{U_a}{2} n_{i,a}(n_{i,a} - 1) + V_{i,a} n_{i,a} \right] \\ &- t_b \sum_{\langle i,j \rangle} b_i^\dagger b_j + \sum_i \left[\frac{U_b}{2} n_{i,b}(n_{i,b} - 1) + V_{i,b} n_{i,b} \right] \\ &+ U_{ab} \sum_i n_{i,a} n_{i,b} + \text{H.c.} \end{aligned} \quad (4.1)$$

where $n_{i,\alpha} \overset{\text{def.}}{\equiv} \alpha_i^\dagger \alpha_i$. The t_α's describe hopping rates and $U_{\alpha\beta}$ the interaction where α and β label the species ($|a\rangle$ or $|b\rangle$). We abbreviate $U_\alpha = U_{\alpha\alpha}$. We have absorbed the chemical potentials and hyperfine splittings into the trapping potential, writing $V_{i,\alpha} = \mathcal{V}_{i,\alpha} + \epsilon_\alpha^{\text{HF}} - \mu_\alpha$ where $\mathcal{V}_{i,\alpha}$ is the external potential at site i for species α, and $\epsilon_b^{\text{HF}} - \epsilon_a^{\text{HF}} = \Delta_0$ is the vacuum hyperfine splitting. In terms of microscopic quantities, the Hubbard parameters are $t_\alpha = \int d\mathbf{r}\, w_\alpha^*(\mathbf{r})\left[-\hbar^2/(2m_\alpha)\nabla^2 + V_{\text{per}}(\mathbf{r})\right] w_\alpha(\mathbf{r})$, $\mathcal{V}_{i,\alpha} \approx V_{\text{trap}}(\mathbf{R_i})$, and $U_{\alpha\beta} = \left(4\pi\hbar^2 a_{\alpha\beta}/m\right) \int d\mathbf{r}\ |w_\alpha(\mathbf{r})|^2 |w_\beta(\mathbf{r})|^2$where m_α is the mass and w_α the normalized Wannier function for state α, while $a_{\alpha\beta}$ denotes the $\alpha - \beta$ scattering length. For ^{87}Rb, the relevant scattering lengths are $a_{aa} = 5.32$ nm and a_{ab}=5.19 nm [12]. The $|b\rangle$ atoms will be sufficiently dilute that a_{bb} will not enter our calculation. The competition between the kinetic and interaction terms drives the Mott insulator to superfluid phase transition.

In the experiments of interest, the atoms all begin in the $|a\rangle$ state, and one measures the rate at which atoms are transferred to the $|b\rangle$ state under the influence of a weak probe of the form $H_{\text{probe}} \propto \sum_j b_j^\dagger a_j e^{-i(\omega+\mu_b-\mu_a)t} + \text{H.c.}$, within the rotating wave approximation, where ω is the frequency of the photons.

(The factors of μ_b and μ_a arise from the canonical transformation which introduces the chemical potential into the Hamiltonian.) To calculate this response, it is sufficient to understand the properties of the single-component Bose–Hubbard model (the terms in (4.1) containing only a's).

Formally the lack of $|b\rangle$ atoms in the initial state implies that $\mu_b < \epsilon(0)$ where $\epsilon(0)$ is the energy of the $k = 0$ single particle state. All (zero-temperature) observables will be independent of μ_b.

Mean-field theory. The ground state of the single component Bose–Hubbard model is well approximated by the Gutzwiller mean-field theory (GMF) of Ref [13]. This approach is exact in infinite dimensions and in the deep Mott insulator and superfluid limits. Sophisticated numerical calculations, some with a trapping potential, have shown that this mean field theory yields qualitatively accurate phase diagrams, energies, and spatial density profiles [14–18]. As a point of reference, Monte-Carlo calculations predict that for unity filling the 3D Bose–Hubbard model on a cubic lattice has an insulator–superfluid transition at $t/U = 0.03408(2)$, while mean field theory gives $t/U = 0.029$ [19]. We will work within this approximation. As will be apparent, one could extend our results to include fluctuation effects by numerically calculating the density and compressibility of the homogeneous system. Within the local density approximation discussed below these homogeneous quantities are the only theoretical input needed to determine the spectrum.

The Gutzwiller mean field approximation to the Bose–Hubbard model can be developed either from a mean-field or variational standpoint. As a variational ansatz, GMF corresponds to taking the wavefunction to be a tensor product of states at each site: $|\Psi\rangle = \bigotimes_i \left(\sum_n f_n^{(i)} |i, n\rangle\right)$ where $|i, n\rangle$ is the state with n particles at the i'th site; the $f_n^{(i)}$ are varied. In the corresponding mean-field language, fluctuations of the annihilation operators from their expectations are assumed not to affect neighboring sites. Then, assuming translation symmetry remains unbroken and letting q be the number of nearest neighbors, one has

$$H_{\mathrm{MF}} = \sum_i \left[-qta_i^\dagger \langle a \rangle + U\frac{n_i^2}{2} + V_i n_i + \mathrm{H.c.} \right] \tag{4.2}$$

from which one can find a self-consistent ground state with $\langle a \rangle = \sum_n \sqrt{n+1} f_n^* f_{n+1}$.

As a mean-field theory, this approach cannot accurately determine some of the properties of the system near the critical point at the tip of the Mott lobe. Furthermore, it treats the Mott state as inert, neglecting small, but finite, density fluctuations. Given the smallness of these fluctuations in three dimensions [20], we do not believe that they play an important role in the experiments. Furthermore, whatever role they do play will likely be obscured by the trap.

Local density approximation. We use a local density (or Thomas–Fermi) approximation (LDA) to calculate the spatial dependence of thermodynamic quantities: all physical quantities at location $\mathbf{r}$ are taken to be those of a homogeneous system at a chemical potential $\mu - \mathcal{V}\mathbf{r})$. This is expected to be valid when the spatial correlation length of the homogeneous system is much smaller than the length scale of the trapping potential [21]. The validity of the GMF + LDA is discussed in Refs. [14–18].

4.3.2 Homogeneous Clock Shifts

The clock shift is a density-dependent shift in the energy splitting Δ for driving a transition from internal atomic states $|a\rangle$ to states $|b\rangle$ due to the inter-particle interactions. In this section we review the known results for the clock shift of a homogeneous system in terms of local correlations, and specialize to the case of atoms in a periodic potential.

We will assume that $t_a = t_b$ and $\mathcal{V}_{i,b} = \mathcal{V}_{i,a}$. These assumptions are justified in the recent experiments, where the polarizabilities of the two internal states are nearly indistinguishable. In linear response, the average clock shift energy of the homogeneous system, is then

$$\delta E_2 = (U_{ab} - U_\alpha)\frac{\langle\sum_i a_i^\dagger a_i^\dagger a_i a_i\rangle}{\langle\sum_i a_i^\dagger a_i\rangle} \tag{4.3}$$

where the expectation is in the initial, all-$|a\rangle$ state [22, 23]. This expression can be rewritten in a somewhat more familiar form by defining the local second order correlation function $g_2 \stackrel{\text{def.}}{\equiv} \langle a_i^\dagger a_i^\dagger a_i a_i\rangle / \langle a_i^\dagger a_i\rangle^2$ so

$$\delta E_2 = (U_{ab} - U_\alpha) g_2 \langle n\rangle. \tag{4.4}$$

Special cases of the clock-shift formula: dilute superfluid, Mott insulator, and normal fluid. For a dilute superfluid, the initial state is a coherent state, and (4.3) gives a shift proportional to the occupation of each site,

$$\delta E_{\mathrm{SF}} = (U_{ab} - U_a)n. \tag{4.5}$$

Deep within the Mott insulating phase, the initial state is a number eigenstate and

$$\delta E_{\mathrm{MI}} = (U_{ab} - U_a)(n-1).$$

This latter formula has an intuitive explanation. In a Mott insulator with filling of one particle per site, the atoms are isolated so there is no interaction between particles. Hence δE_{MI} must vanish when $n = 1$. Figure 4.2 illustrates how the clock shift energy evolves from being proportional to n to $n - 1$ by juxtaposing the contours of fixed δE_2 and those of fixed density.

If one raises the temperature the system becomes a normal fluid, even at weak interactions. In the absence of interactions, the normal fluid statistical factor

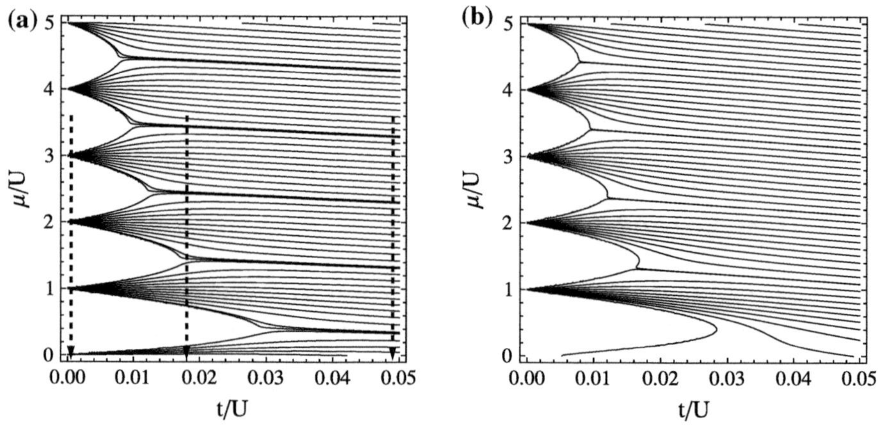

Fig. 4.2 **a** Gutzwiller Mean Field phase diagram with constant density contours. The *vertical dashed lines* show the spatial dependance of the chemical potential for a trapped gas in the LDA: from left-to-right these correspond to the deep Mott limit, the "corrugated superfluid" situation appropriate to Fig. 4.4b, and the dilute superfluid. **b** Phase diagram with contours of constant $\langle n\rangle g_2 = \langle n(n-1)\rangle/\langle n\rangle$. Contours are spaced by 0.1, with additional lines at $m \pm 0.01$, for integer m, to emphasize the Mott regions

g_2 appearing in (4.4) is $g_2 = 2$, [24] so that the clock shift energy is twice as large as in the superfluid:

$$\delta E_{\text{NF}} = 2(U_{ab} - U_a)n. \tag{4.6}$$

Given that there is no phase transition between the zero temperature Mott insulator and the normal gas, it is interesting that the clock shift energy changes from $2(U_{ab} - U_a)n$ in the normal fluid to $(U_{ab} - U_a)(n-1)$ in the Mott insulator. A quantitative understanding of this crossover would require calculating the temperature dependence of the pair correlations in the strongly interacting limit.

4.3.3 Calculation of Spectrum in a Trap

To calculate the spectrum we assume that the gas can be treated as locally homogeneous, and we can independently sum the spectrum from each region in the cloud. The number of atoms of atoms transferred to the $|b\rangle$ state will be

$$\begin{aligned} N_b(\omega) &\propto \int d^3r\, n(r) I_{\mathbf{r}}(\omega) \\ &\approx \int d^3r\, n(r)\delta_{1/\tau}(\Delta(\mathbf{r}) - \omega), \end{aligned} \tag{4.7}$$

where $I_{\mathbf{r}}(\omega)$ is the spectrum of a homogeneous system with chemical potential $\mu(\mathbf{r}) = \mu_0 - V_{\text{trap}}(\mathbf{r})$ appropriate to position $\mathbf{r}$ in the trap. Averaging over the trap will wash-out all of the structure in $I_{\mathbf{r}}(\omega)$, so we make the simplifying approximation of replacing it with a broadened delta function $\delta_{1/\tau}(\omega - \Delta(\mathbf{r}))$, where the peak location, $\Delta(\mathbf{r})$, is given by the mean clock-shift of a homogeneous system with chemical potential $\mu(\mathbf{r}) = \mu_0 - V_{\text{trap}}(\mathbf{r})$. We will model $\delta_\gamma(v) = (1/\pi)\gamma/(v^2 + \gamma^2)$ as a Lorentzian of width γ, and use $\tau = 1/\gamma = 100$ ms, corresponding to the finite probe duration in the experiments.

Note that the replacement of $I_{\mathbf{r}}(\omega)$ by $\delta_{1/\tau}(\omega - \Delta(\mathbf{r}))$ is a severe approximation. The homogeneous spectrum can have important structure, even displaying bimodality in a narrow parameter range. However, inhomogeneous broadening obscures this structure, and we believe that one can adequately model the experiment via (4.7).

We calculate the integral in (4.7) within the Gutzwiller mean field approximation to the Bose–Hubbard model. As illustrated in Fig. 4.2, both the density n and the clock shift Δ can be expressed as functions of the parameters μ/U and t/U. Within the local density approximation, t is constant throughout the trap, and μ varies in space, taking its maximal value μ_0 at the center of the trap.

For extreme values of t/U (either large or small) we can analytically calculate the contours in Fig. 4.2a Generically, however we must rely on numerical methods.

Our results are shown in Fig. 4.1 for a harmonic trap $V_{\text{trap}}(\mathbf{r}) = m\omega_x^2 x^2/2 + m\omega_y^2 y^2/2 + m\omega_z^2 z^2/2$. One sees that in the deep Mott limit, the spectrum displays sharp peaks, while in the deep superfluid limit, the lineshape is smooth. The peaks are due to the stepwise variation of $\Delta(\mu)$ (illustrated in Fig. 4.2a) which lead to large regions of the trap where $\Delta(r)$ takes on discrete values. Compared with the experiments in Ref. [8], our spectral lines have severely reduced small detuning spectral weight. In Sect. 4.4 we show that trap anharmonicities can account for this difference.

Note that within the local density approximation, the spectrum is independent of trap anisotropies, as long as the trap is harmonic. This generic feature of the LDA is seen by examining an integral of the form $I = \int d^3r f(\mu) = \int d^3r f(\mu_0 - m\omega_x^2 x^2/2 - m\omega_y^2 y^2/2 - m\omega_z^2 z^2/2)$. Rescaling the coordinates so that $m\omega_x^2 x^2/2 = \mu_0 \bar{x}^2$ (and similarly for y and z), this integral becomes $I = \sqrt{8\mu_0^3/m^3\omega_x^2\omega_y^2\omega_z^2} 4\pi \int d\bar{r}\, \bar{r}^2 f(\mu_0 - \mu_0\bar{r}^2)$, where $\bar{r} = \sqrt{\bar{x}^2 + \bar{y}^2 + \bar{z}^2}$. From this analysis it is clear that apart from an overall scale factor, the spectral lineshape is only a function of the central chemical potential μ_0.

Experimental Parameters. The experimental control parameters are the optical lattice depth V_0, the number of particles N, and the trap frequencies ω_v. The natural theoretical parameters are t, U, and μ_0. To compare our results to experiment, we use a non-interacting band structure calculation to relate t and U to V_0 [11]. To relate μ_0 to experimental parameters we note that within the LDA the number of trapped atoms N is only a function of t/U, and $\mu_0/\hbar\bar{\omega}$ where

$\bar{\omega}^3 = \omega_x\omega_y\omega_z$. For each value of t/U we compute $N(\mu_0/\hbar\bar{\omega})$, for several values of μ_0, then invert the function to get μ_0 as a function of N. We then have the ability to select the value of μ_0 corresponding to the number of particles used in the experiment.

Campbell et al. [8] do not report the number of particles in the experiment. For Fig. 4.1 , we choose $N = 9 \times 10^4$ so that the maximum site filling for $V_0 = 35E_{\mathrm{rec}}$ and $V_0 = 25E_{\mathrm{rec}}$ is $n = 5$, as is observed in the experiment.

4.3.4 Analytic Results

4.3.4.1 Dilute Superfluid

Having numerically calculated the spectra, we now specialize to the dilute superfluid limit where the line shape can be calculated analytically. The clock shift energy in this limit is $\Delta = (U_{ab} - U_a)\langle n\rangle$, and within the local density approximation the site filling at position $\mathbf{r}$ is the greater of zero and

$$n(r) = (\mu_0 - \epsilon(0) - V_{\mathrm{trap}}(\mathbf{r}))/U_a \tag{4.8}$$

where as previously stated, $V_{\mathrm{trap}}(\mathrm{r}) = m\omega_x x^2/2 + m\omega_y^2/2 + m\omega_z z^2/2$ is the trapping potential, μ_0 is the central chemical potential. In the tight binding limit, the energy of the single-particle $k = 0$ state is $\epsilon(0) = -qt$ where q is the number of nearest neighbors. Substituting this result into (4.7), and neglecting the broadening one finds

$$N_b(\delta\omega) \propto \delta\omega\sqrt{(U_{ab} - U_a)n(0) - \delta\omega}, \tag{4.9}$$

where $n(0) = (\mu_0 - \epsilon(0))/U_a$ is the central density. Similar expressions were obtained in Ref. [25].

At fixed central chemical potential (equivalently, fixed central density) the width of the spectrum is proportional to $U_{ab} - U_a$, and hence U. Given a fixed number of particles, the central density varies as $n(0) \sim 1/U^{3/5}$, so the width of the spectral line instead varies as $Un(0) \sim U^{2/5}$.

4.3.4.2 Deep Mott Limit

Now we analytically calculate the spectrum in the deep Mott insulator limit, where the density of the homogeneous system with chemical potential μ equals the smallest integer exceeding μ/U, denoted $\lceil \mu/U \rceil$ [11, 18]. In the local density approximation the density jumps from density $n - 1$ to n as one moves through the location in space where local chemical potential is given by $\tilde{\mu} = U_a(n - 1)$. Each plateau of fixed n gives a (broadened) delta-function contribution to the line shape

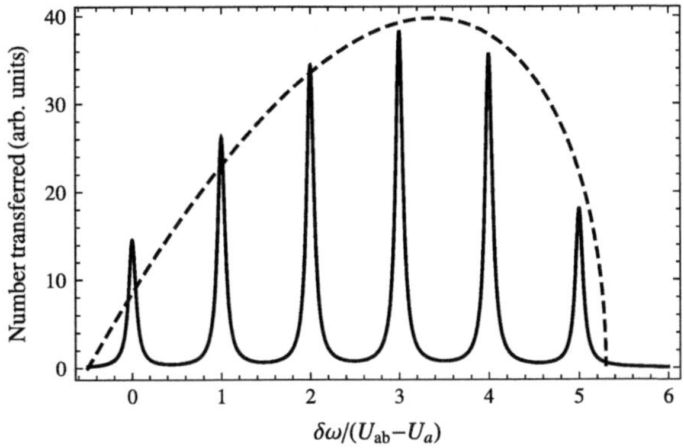

Fig. 4.3 Analytically calculated spectra for the harmonically trapped system in the deep Mott limit (*solid line*), plotted as a function of $\delta\omega/(U_{ab} - U_a)$. Superimposed is the spectrum of the superfluid (*dashed line*) with the same parameters, but horizontally shifted to the left by $-(\delta\omega/2)(U_{ab} - U_a)$. The central density is $n_s(0) = [\mu_0 - \epsilon(0)]/U_a = 5.8$. This illustrates that the envelope of the spectral line in the Mott insulating state has the same shape as the superfluid spectrum, shifted horizontally

at detuning $\delta\omega_n = (n-1)(U_{ab} - U_a)$. The magnitude of the delta function is proportional to the number of particles in the plateau, leading to a spectrum

$$\begin{aligned} N_b(\delta\omega) &\propto \sum_{n=1}^{\bar{n}} A_n \delta_{1/\tau}(\delta\omega - \delta\omega_n) \\ A_{\bar{n}} &= [\mu_0 - U_a(\bar{n}-1)]^{3/2}\bar{n} \\ A_{n\neq\bar{n}} &= \left[(\mu_0 - U_a(n-1))^{3/2} - (\mu_0 - U_a n)^{3/2}\right] n, \end{aligned} \tag{4.10}$$

where $\bar{n} = \lceil \mu_0/U_a \rceil$ is the central density.

The deep superfluid and deep Mott insulator spectra are plotted in Fig. 4.3 using Eqs. 4.9 and 4.10. Note the envelope of the spikes seen in the insulating state has the same shape as the superfluid spectrum. This can be understood from noting that in both cases the density is proportional to μ or $\lceil \mu/U \rceil$, resulting in similar coarse-grained Δ.

4.3.5 Intermediate Coupling

Finally, let's consider how the spectrum evolves as one increases t/U from zero. For non-zero t/U, superfluid shells form between Mott plateaus. These regions make the density continuous. Consequently, in the spectra, the areas of zero signal between peaks begin to fill in. Using our numerics, we find that the peaks remain visible until the system is well into the superfluid regime. An example is shown in

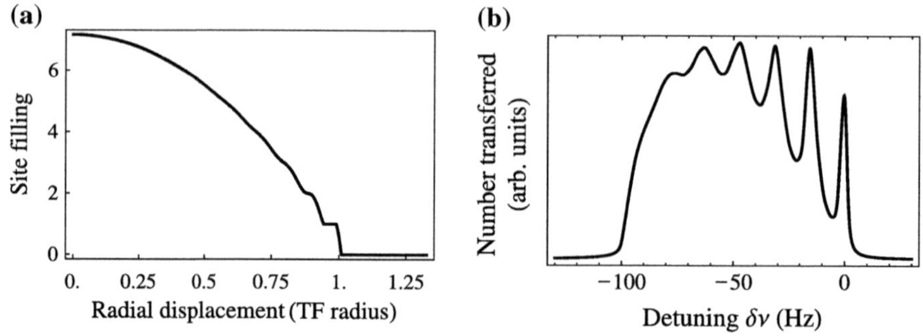

Fig. 4.4 **a** The density as a function of distance from the trap center for a harmonic trap in units of the Thomas–Fermi radius ℓ_{TF}, defined as $\ell_{TF} \equiv \sqrt{(\mu_0 - \epsilon_0)/(m\omega^2/2)}$. The density profile corresponds to the LDA contour at $t/U_a = 0.018$ of Fig. 4.2a. The corrugation of the density is observable, but not dramatic; it would be particularly difficult to image in experiments looking at columnar integrated densities. **b** The hyperfine spectra arising from the density plotted in (**a**)

Fig. 4.4b, corresponding to the chemical potential trajectory at $t/U_a = 0.018$ shown in Fig. 4.2a. Note that although the only Mott lobe crossed is at $n = 1$, six peaks are clearly visible. Clearly one must be cautious about using such spectra to distinguish superfluid and Mott insulating states.

The source of the peaks for $t/U_a = 0.018$ is weak density corrugations which arise in the *superfluid* state near the Mott boundaries. These corrugations can be inferred from the unequal spacing of the isodensity contours in Fig. 4.2. The spectrum is a powerful amplifier of these corrugations, as they are hardly prominent in the real-space density shown in Fig. 4.4a.

4.4 Refinements

As seen previously, GMF + LDA captures the main features of the experimental spectra: sharp peaks occur in the Mott insulator limit, a smooth distribution in the deep superfluid limit. Furthermore, the overall energy scales of our spectra are consistent with those found experimentally. We caution however that we have treated the harmonic trapping frequency as independent of the lattice depth, while experimentally the harmonic confinement varies in an uncharacterized manner when the optical lattice intensity is changed. With this systematic variation makes quantitative comparison difficult.

Despite the qualitative similarities between theory and experiment, serious discrepancies remain. In particular, the experiment finds much more spectral weight at small detunings than theory predicts. Here we explore possible sources of this discrepancy. Our primary result is that the discrepencies are consistent with trap anharmonicities.

In Sect. 4.4.1 we give an analysis of trap anharmonicities. In the following sections we briefly discuss several other possible explanations of the discrepencies: non-equilibrium effects and nonlinearities in the transfer rate. Although these latter two effects could distort the spectrum in a manner qualitatively consistent with experiment, we find that neither of them plays a significant role in these particular experiments.

4.4.1 Anharmonicity

The trap used in the experiments of Ref. [8] is a combination of an Ioffe–Pritchard magnetic trap, which is roughly harmonic, and an optical trap, which provide highly anisotropic Gaussian confinement. This results in a trap with "soft" anharmonicities, increasing the number of particles in the low density tails of the cloud. This will accentuate the small $\delta\omega$ peaks in the spectrum. The presence of anharmonicities is clear in Fig. 4.4 of Ref. [8], where the spatial distribution of the Mott insulator shells is far from elliptical.

We model the trapping potential as

$$\begin{aligned} V_{\text{trap}}(x,y,z) &= \frac{m\omega_a^2}{2}x^2 + \frac{m\omega_r^2}{2}(y^2+z^2) \\ &\quad + I_a\left(1-e^{-x^2/(2\sigma^2)}\right) + I_r\left(1-e^{-(y^2+z^2)/(2\sigma^2)}\right) \end{aligned} \tag{4.11}$$

where x lies in the soft "axial" direction while y and z constitute the "radial" directions. The $1/e^2$ beam waist is quoted as 70 μm, corresponding to $\sigma = 35$ μm, however we find spatial profiles closer to experiment from the slightly smaller $\sigma = 28$ μm and use this value throughout. The explicit harmonic terms come from the magnetic trap. The anharmonic Gaussian part has two contributions, $I_{o,j}$ from the optical trap and $\alpha_j V_0$ from the optical lattice inducing a further trapping potential, for some constants α_j, with $I_j = I_{o,j} + \alpha_j V_0$. The parameters $\omega_a, \omega_r, \alpha_a$, and α_r are determined from $I_{o,a}, I_{o,r}$ and the quadratic trap frequencies Ω_{j,V_0} at $V_0 = 0$ and $V_0 = 40E_{\text{rec}}$ by matching the quadratic terms of (4.11), giving

$$\begin{aligned} \alpha_j &= \frac{m\sigma^2}{40E_{\text{rec}}}\left(\Omega_{j,40}^2 - \Omega_{j,0}^2\right), \\ \omega_j^2 &= \Omega_{j,0}^2 - \frac{I_{o,j}}{m\sigma^2}. \end{aligned}$$

The quadratic trap frequencies Ω_{j,V_0} are given in Ref. [8] as $\Omega_{r,0} = 2\pi \times 70$ Hz, $\Omega_{r,40} = 2\pi \times 110$ Hz, $\Omega_{a,0} = 2\pi \times 20$ Hz, and $\Omega_{a,40} = 2\pi \times 30$ Hz. The remaining unknown parameter I_o is chosen to be $I_{o,a} = 1.17E_{\text{rec}}$ so that the spatial density profile appears similar to that in the experiment. We take $I_{o,r} = I_{o,a}$ though $I_{o,r}$ has little effect on the spatial density profiles. This yields $\omega_a = 4.8$ Hz, $\omega_r = 67$ Hz, $\alpha_a = 0.039E_{\text{rec}}^{-1}$, and $\alpha_r = 0.56E_{\text{rec}}^{-1}$ to completely characterize the trapping

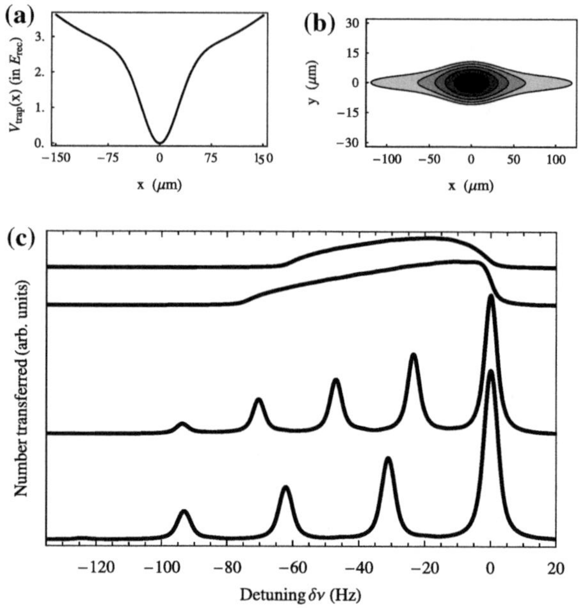

Fig. 4.5 **a** A slice of the anharmonic potential $V_{trap}(\mathbf{r})$ similar to the one found in experiments. **b** Contour lines of constant density in the x–y plane for $n = 1, 2, \ldots, 5$ in the anharmonic trap at $V_0 = 35E_{rec}$. **c** Corresponding spectra (using the "spherical trap model" discussed in text) for $V_0 = 5$, 10, 25, $35E_{rec}$ with $N = 1.4 \times 10^7$ particles

potential of (4.11). Note, that while we have chosen values to mimic the experimental observations, we have not attempted to produce a quantitative "fit" to the experimental data. Figure 4.5b shows the isopotential lines of our model trap.

For numerical efficiency, we produce spectra from a spherically symmetric model with parameters equal to those of the axial direction, which somewhat exaggerates the anharmonic effects. As shown in Fig. 4.5c the small detuning spectral weight is greatly enhanced by the anharmonicity, producing spectra which are consistent with experiments.

4.4.2 *Alternative Explanations of Enhanced Low-Density Spectral Weight*

Here we examine alternative sources of the enhancement of the small detuning spectral weight observed in experiments. We make no claim that this is comprehensive. Rather, this is an examination of the most plausible factors possibly present in the experiments of Ref. [8].

Losses. First, we explore the possibility that three-body collisions drive the cloud out of equilibrium. Atoms are removed preferentially from high density sites, ostensibly enhancing the small-detuning spectral weight. The timescale for decay from the $n = 5$ Mott insulator state is 200 ms [8]. A characteristic equilibration time is the trap period, ~ 10 ms. Given the separation of timescales it is extremely unlikely that the system is far out of equilibrium. Furthermore, the loss rate is effectively zero for one- and two-particle site fillings and hence losses are unable to explain the experimentally observed enhancement of the $n = 1$ peak relative to the $n = 2$ peak.

Nonlinearities in transfer rate. The probes used to measure the spectrum may possibly drive the system out of the linear regime where the transfer rate is proportional to the density. For example, if the transition becomes saturated in the high density regions of the cloud, then the observed spectral weight will be reduced at large detunings. However, the density dependence of these saturation effects is slow, making it unlikely that they could not be responsible for the dramatic suppression of the ratio of the spectral weight in the $n = 2$ and $n = 1$ peaks. A model calculation in the deep Mott regime, where the sites decouple, confirms this result.

Non-equilibrium effects. Another possible mechanism for distorting the spectrum would be that the relatively short lattice ramp time might drive the system out of thermal equilibrium. Modelling these nonequilibrium effects is nontrivial, however we can put an upper bound on the size of the distortion by considering the "sudden limit," where the lattice intensity is quickly increased, quenching the superfluid into the Mott state. After the quench, each decoupled site will be in a quantum superposition of different particle numbers, with a Poisson probability distribution. The spectrum will then consist of a series of discrete peaks with the intensity of the peak at detuning $\nu_m = (m-1)(U_{ab} - U_a)$ given by $I_m = \int d^3r\, n(\mathbf{r})^m e^{-n(\mathbf{r})}/(m-1)!$, where $m = 1, 2, \ldots$ and $n(r) = \max\{(\mu_0 - \epsilon(0) - V_{\text{trap}}(\mathbf{r})/U_a, 0\}$ is the density in the superfluid in the Thomas–Fermi approximation. While this non-equilibrium spectrum is significantly different from the equilibrium spectrum, no strong peak at $m = 1$ appears when the central density is much larger than one.

Moreover, since the characteristic equilibration time is of order of the trap period, ~ 10 ms, while the lattice is changed with a characteristic timescale of 40 ms, we expect the system to be very near equilibrium [8], and any nonequilibrium effects should be much reduced compared to those predicted by the sudden approximation.

4.5 Summary

We calculate the hyperfine spectra of trapped bosonic atoms in an optical lattice. We consider the cases of harmonic and model anharmonic traps. We show that a harmonic trap produces a spectrum which shares qualitative features with the experimental spectra measured by Campbell et al. [8]: in the deep superfluid limit one has a smooth peak, while in the deep Mott limit, one sees several discrete peaks. To reproduce the small-detuning spectral weights, however, trap anharmonicities are necessary.

We find the spectra are extremely sensitive to density corrugation. As an example, the mild density corrugations which are found in the superfluid near the

Mott insulator boundary are sufficient to produce a pronounced splitting of the spectral line. Consequently, the spectra are continuous across the superfluid to Mott insulator transition. Such continuity is characteristic of a second-order phase transition, and makes identifying the superfluid transition difficult.

We acknowledge an illuminating discussion with Jim Sethna regarding the Mott insulating state's clock shift energy. We thank Wolfgang Ketterle and Gretchen Campbell for information regarding their experiment, and Smitha Vishveshwara for critical comments. This work was supported by NSF grant No. PHY-0456261 and GAANN Award No. P200A030111 from the US Department of Education.

References

1. G.K. Brennen, G. Pupillo, A.M. Rey, C.W. Clark, C.J. Williams, J. Phys. B At. Mol. Opt. Phys. **38**, 1687 (2005)
2. P. Rabl, A.J. Daley, P.O. Fedichev, J.I. Cirac, P. Zoller, Phys. Rev. Lett. **91**, 110403 (2003)
3. M. Greiner, O. Mandel, T. Esslinger, T.W. Hänsch, I. Bloch, Nature **415**, 39 (2002)
4. S. Fölling, F. Gerbier, A. Widera, O. Mandel, T. Gericke, I. Bloch, Nature **434**, 481 (2005)
5. S. Fölling, A. Widera, T. Muller, F. Gerbier, I. Bloch, Phys. Rev. Lett. **97**, 060403 (2006)
6. M. Greiner, O. Mandel, T.W. Hänsch, I. Bloch, Nature **419**, 51–54 (2002)
7. I. Bloch, J. Dalibard, W. Zwerger, Rev. Mod. Phys. **80**, 885 (2008)
8. G.K. Campbell, J. Mun, M. Boyd, P. Medley, A.E. Leanhardt, L.G. Marcassa, D.E. Pritchard, W. Ketterle, Science **313**, 649 (2006)
9. D.G. Fried, T.C. Killian, L. Willmann, D. Landhuis, S.C. Moss, D. Kleppner, T.J. Greytak, Phys. Rev. Lett. **81**, 3811 (1998)
10. M.R. Matthews, D.S. Hall, D.S. Jin, J.R. Ensher, C.E. Wieman, E.A. Cornell, F. Dalfovo, C. Minniti, S. Stringari, Phys. Rev. Lett. **81**, 243 (1998)
11. D. Jaksch, C. Bruder, J.I. Cirac, C.W. Gardiner, P. Zoller, Phys. Rev. Lett. **81**, 3108 (1998)
12. van E.G.M. Kempen, S.J.J.M.F. Kokkelmans, D.J. Heinzen, B.J. Verhaar, Phys. Rev. Lett. **88**, 093201 (2002)
13. M.P.A. Fisher, P.B. Weichman, G. Grinstein, D.S. Fisher, Phys. Rev. B **40**, 546570 (1989)
14. L. Pollet, S. Rombouts, K. Heyde, J. Dukelsky, Phys. Rev. A **69**, 043601 (2004)
15. S. Bergkvist, P. Henelius, A. Rosengren, Phys. Rev. A **70**, 053601 (2004)
16. S. Wessel, F. Alet, M. Troyer, G.G. Batrouni, Phys. Rev. A **70**, 053615 (2004)
17. G.G. Batrouni, V. Rousseau, R.T. Scalettar, M. Rigol, A. Muramatsu, P.J.H. Denteneer, M. Troyer, Phys. Rev. Lett. **89**, 117203 (2002)
18. B. DeMarco, C. Lannert, S. Vishveshwara, T. Wei, Phys. Rev. A **71**, 063601 (2005)
19. B. Capogrosso-Sansone, N.V. Prokof'ev, B.V. Svistunov, Phys. Rev. B **75**, 134302 (2007)
20. B. Capogrosso-Sansone, E. Kozik, N. Prokof'ev, B. Svistunov, Phys. Rev. A **75**, 013619 (2007)
21. C.J. Pethick, H. Smith, Bose–Einstein condensation in dilute gases. (Cambridge University Press, Cambridge, 2001)
22. M. Öktel, T.C. Killian, D. Kleppner, L.S. Levitov, Phys. Rev. A **65**, 033617 (2002)
23. C.J. Pethick, H.T.C. Stoof, Phys. Rev. A **64**, 013618 (2001)
24. L. Landau, E. Lifshitz, Statistical Physics I. (Butterworth Heinemann, London, 1997)
25. J. Stenger, S. Inouye, A.P. Chikkatur, D.M. Stamper-Kurn, D.E. Pritchard, W. Ketterle, Phys. Rev. Lett. **82**, 4569 (1999)

Chapter 5
RF Spectra: Multiple Peaked Spectrum in a Homogeneous System

This chapter will discuss RF spectroscopy of lattice bosons from a point of view complementary to that of Chap. 4. That chapter looked at things from the vantage of sum rules assuming a single, narrow spectral line, one common approach in the literature. This chapter is more closely related to another common approach in the literature: taking the RF spectra to be (convolutions of) the single particle spectral functions, perhaps with shifts of lines due to final state interactions. This naturally leads to a two-peaked structure in the superfluid: there are gapless Goldstone modes (superfluid phonons) and gapped order parameter amplitude modes (Higgs modes). After constructing an approximation sufficient to capture the global physics, we will reconcile the two points of view. In addition to giving a global calculation for the spectrum of bosons in an optical lattice, the reconciliation provides a new general lesson for interpreting RF spectra of many-body systems, showing how standard viewpoints emerge in special limits of the proper calculation.

5.1 Chapter Overview

We calculate the radio-frequency spectrum of a trapped cloud of cold bosonic atoms in an optical lattice. Using random phase and local density approximations we produce both trap averaged and spatially resolved spectra, identifying simple features in the spectra that reveal information about both superfluidity and correlations. Our approach is exact in the deep Mott limit and in the dilute superfluid when the hopping rates for the two internal spin states are equal. It contains final

This chapter was adapted from "Many-body physics in the radio-frequency spectrum of lattice bosons" by Kaden. R. A. Hazzard and Erich. J. Mueller, published in Physical Review A **81**, 033404 (2010).

K. R. A. Hazzard, *Quantum Phase Transitions in Cold Atoms and Low Temperature Solids*, Springer Theses, DOI: 10.1007/978-1-4419-8179-0_5,

state interactions, obeys the Ward identities (and the associated conservation laws), and satisfies the f-sum rule. Motivated by earlier work by Sun et al. (Phys Rev A **79**, 043422, 2009), we also discuss the features which arise in a spin-dependent optical lattice.

5.2 Introduction

Bosonic atoms in optical lattices, described by the Bose–Hubbard model [1, 2], display a non-trivial quantum phase transition between a superfluid and Mott insulator. The latter is an incompressible state with an integer number of atoms per site. In a trap the phase diagram is revealed by the spatial structure of the gas: one has concentric superfluid and insulating shells. This structure has been elegantly explored by radio frequency (RF) spectroscopy [3], a technique which has also given insight into strongly interacting Fermi gases across the BEC–BCS crossover [4]. Here we use a Random Phase Approximation (RPA) that treats fluctuations around the strong coupling Gutzwiller mean field theory to explore the radio-frequency spectrum of lattice bosons.

We find two key results: (1) Our previous sum-rule based analysis [5] of experiments at MIT [3] stands up to more rigorous analysis: in the limit of small spectral shifts, the RPA calculation reduces to that simpler theory. (2) In a gas with more disparate initial and final state interactions (such as Cesium), the spectrum becomes more complex, with a bimodal spectrum appearing even in a homogeneous gas. The bimodality reveals key features of the many-body state. For example, in the limit considered by Sun et al. [6], the spectral features are related to the nearest-neighbor phase coherence. In the Gutzwiller approximation, the phase coherence directly maps onto the condensate density. In this paper we provide a physical picture of this result and explain how this bimodality can be observed in a spatially resolved experiment.

5.2.1 *RF Spectroscopy*

In RF spectroscopy, a radio wave is used to flip the hyperfine spin of an atom from $|a\rangle$ to $|b\rangle$. The rate of excitation reveals details about the many-body state because the $|a\rangle$ and $|b\rangle$ atoms have slightly different interactions. Generically the interaction Hamiltonian is $H_{\text{int}} = \sum_j U_{aa}n_a(n_a - 1)/2 + U_{bb}n_b(n_b - 1)/2 + U_{ab}n_an_b$, with $U_{aa} \neq U_{ab} \neq U_{bb}$, where n_σ is the number of σ-state atoms on site j. In the simplest mean-field picture, the energy needed to flip an atom on site j from state a to state b is shifted by an energy $\delta\omega = U_{bb}n_b + (U_{ab} - U_{aa})n_a$. Applying this picture to an inhomogeneous gas suggests that the absorption spectrum reveals a histogram of the atomic density. Such a density probe is quite valuable:

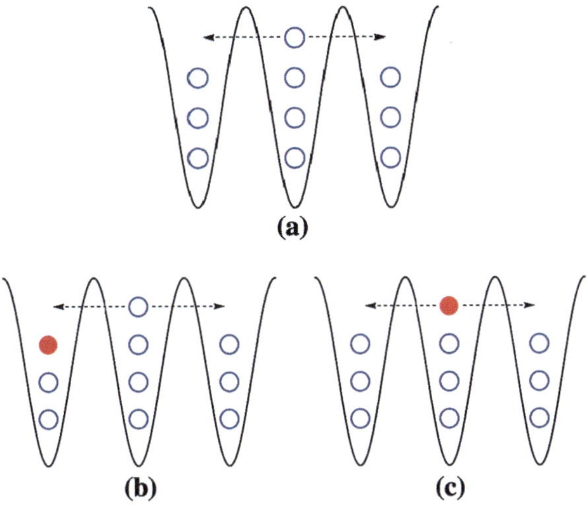

Fig. 5.1 (Color online) Illustration of two types of RF-active excitations of the lattice superfluid near the Mott transition. *Open (blue) circles* are atoms in the $|a\rangle$ state, *filled (red) circles* are atoms in the $|b\rangle$ state, and the *arrows* indicate a delocalized particle while other particles are localized. **a** Illustrates the initial superfluid state, consisting of a dilute gas of atoms moving in a Mott background. Final states in **b** and **c**, show the excitation of a core or delocalized atom

in addition to the aforementioned examples, it was the primary means of identifying Bose–Einstein condensation in atomic hydrogen [7].

Recently Sun et al. [6] found a bimodal spectrum in a special limit of this problem, as did Ohashi et al. [8] in a separate limit, calling into question this simple picture. We give a simple physical interpretation of the bimodality. As illustrated in Fig. 5.1, the superfluid state near the Mott insulator can be caricatured as a dilute gas of atoms/holes moving in a Mott background. An RF photon can either flip the spin of one of the core atoms, or flip the spin of one of the mobile atoms. The energy of these two excitations will be very different, implying that the RF spectrum should be bimodal. Through our RPA calculation, we verify this feature, calculating the frequencies of the two peaks and their spectral weights. Interestingly, this calculation reveals that the two excitations in our cartoon are strongly hybridized.

We find that for parameters relevant to previous ^{87}Rb experiments, the degree of bimodality is vanishingly small and our previous sum rule arguments [5] accurately describe such experiments. On the other hand, there are opportunities to study other atoms (for example, Na, Cs, Yb) for which the bimodality may be more pronounced. Moreover, if the interactions or tunneling rates can be tuned via a spin-dependent lattice or a Feshbach resonance then this spectral feature will appear in a dramatic fashion.

This bimodal spectrum, with one peak produced by the "Mott" component and another by the "superfluid" component, is reminiscent of the spectrum of a finite temperature Bose gas in the absence of a lattice. As described by Oktel and Levitov [9], in that situation one sees one peak from the condensate, and one from the incoherent thermal atoms. We would expect that at finite temperature our "Mott" peak continuously evolves into their "thermal" peak.

5.3 Bose–Hubbard Model

5.3.1 Model and RF Spectra

In the RF spectra experiments we consider, initially all atoms are in the a-internal state and the RF pulse drives them to the b-state. Consequently, we consider two-component bosons trapped in the periodic potential formed by interfering laser beams, described by a Bose–Hubbard model [1],

$$H = -\sum_{\substack{\langle i,j\rangle \\ \sigma=\{a,b\}}} t_\sigma c_{i,\sigma}^\dagger c_{j,\sigma} + \sum_{\sigma,j}(V_{j,\sigma} - \mu_\sigma)c_{j,\sigma}^\dagger c_{j,\sigma} + \sum_j \left(\sum_{\alpha,\beta} \frac{U_{\alpha\beta}}{2} c_{j,\alpha}^\dagger c_{j,\beta}^\dagger c_{j,\beta} c_{j,\alpha}\right), \tag{5.1}$$

where c_σ and $c_\sigma^\dagger$ are the annihilation and creation operators for states in the internal state σ, μ_σ is the chemical potential, $V_{j,\sigma}$ is the external potential with δ, the vacuum $a-b$ splitting, absorbed into it, $U_{\alpha\beta}$ is the α state-β state on-site interaction strength, and t_σ is the hopping matrix element. The interactions are tunable via Feshbach resonances and spin-dependent lattices are also available [10]. For this latter setup, the hopping matrix elements may be tuned by the intensity of the lattices, and introducing small displacements of the lattice will reduce the overlap between the Wannier states of a and b atoms, and therefore may also be an efficient way to control the relative size of U_{aa} and U_{ab}. The interaction U_{bb} will be irrelevant: we will only consider the case where there is a vanishingly small concentration of b-state particles. In calculating the response to RF photons we will take $V_j =$ constant. Trap effects will later be included through a local density approximation [5] which is valid for slowly varying traps [8, 11–19].

Experimentally the RF spectrum is measured by counting the number of atoms transferred from state a to b when the system is illuminated by a RF pulse. These dynamics are driven by a perturbation

$$H_{\mathrm{rf}} = \sum_j \gamma(t) c_{j,b}^\dagger c_{j,a} + \mathrm{H.c.} \tag{5.2}$$

where $\gamma(t)$ is proportional to the time-dependent amplitude of the applied RF field multiplied by the dipole matrix element between states a and b: typically γ is a sinusoidal pulse with frequency ω with a slowly varying envelope ensuring a small bandwidth. Due to the small wave-number of RF photons, recoil can be neglected.

For a purely sinusoidal drive, Fermi's Golden Rule gives the number of atoms transferred per unit time for short times to be

$$\Gamma(\omega) = \frac{2\pi}{\hbar} \sum_{i,f} p_i \delta(\omega - (E_f - E_i)) |\langle f|H_{\mathrm{rf}}|i\rangle|^2 \tag{5.3}$$

where the sum is over the initial states (occupied with probability $p_i = e^{-\beta E_i}$) and the final states, all of which are eigenstates of H with energies E_i and E_f. We will restrict ourselves to $T = 0$ and the physically relevant case where the initial states contain no b-atoms.

5.3.2 Sum Rules

Taking moments of Eq. 5.3 [9, 20], the mean absorbed photon frequency is

$$\langle \omega \rangle = \frac{\int d\omega\, \omega \Gamma(\omega)}{\int d\omega\, \Gamma(\omega)} = \frac{\langle [H_{\rm rf}, H] H_{\rm rf} \rangle}{\langle H_{\rm rf}^2 \rangle} \tag{5.4}$$

$$= \delta - z(t_b - t_a) f_c + (U_{ab} - U_{aa}) g_2 \langle n \rangle. \tag{5.5}$$

We defined δ to be the vacuum $a - b$ splitting, the local phase coherence factor is

$$f_c = \frac{\left\langle c_{i,a}^\dagger c_{j,a} \right\rangle}{\langle n \rangle}, \tag{5.6}$$

with i and j nearest neighbors, the site filling is $n \equiv c_a^\dagger c_a$, and the lattice coordination is z. The zero-distance density–density correlation function is

$$g_2 = \frac{\langle c_a^\dagger c_a^\dagger c_a c_a \rangle}{\langle n \rangle^2}. \tag{5.7}$$

The second term in Eq. 5.5 may be interpreted as the mean shift in the kinetic energy when the spin of an atom is flipped. In particular, within a strong-coupling mean-field picture $\langle c_{i,a}^\dagger c_{j,a} \rangle = \langle c_{i,a}^\dagger \rangle \langle c_{j,a} \rangle$ is the condensate density, which can therefore be measured with this technique. The second term in Eq. 5.5 is the shift in the interaction energy.

Our subsequent approximations will satisfy this sum rule. This is non-trivial: for example, even in simultaneous limits of $t_b = 0$, $U_{ab} = U_{aa}$, and $t_a \to 0$ considered in Ref. [6], their results violate this sum rule by a factor of ~ 3.

Since it plays no role in the remainder of the discussion, we will set to zero the vacuum level splitting: $\delta = 0$. This amounts to working in a "rotating frame".

5.4 Random Phase Approximation

5.4.1 General Setup and Solution

To calculate the RF spectrum we employ a time-dependent strong-coupling mean-field theory which includes $k = 0$ fluctuations around the static strong-coupling Gutzwiller mean field theory [2]. This mean field theory is exact in the deep Mott limit and in the dilute superfluid when $t_a = t_b$, and it yields fairly accurate ground states in the intermediate regime [11–15]. Refs. [8, 19] previously used analogous RPA's to calculate the Bose–Hubbard model's quasiparticle spectra and RF spectra with $U_{ab} = 0$, which reduce to the $k = 0$ single particle spectra.

We use the homogeneous time-dependent Gutzwiller variational ansatz

$$|\psi(t)\rangle = \bigotimes_i \left[\sum_n (f_n(t)|n,0\rangle_i + g_n(t)|n-1,1\rangle_i) \right] \tag{5.8}$$

where $|n_a, n_b\rangle_i$ is the state at site i with n_a particles in the a state and n_b in the b state. The equation of motion for $f_n(t)$ and $g_n(t)$ are derived by minimizing the action $S = \int dt \mathcal{L}$, with Lagrangian

$$\mathcal{L} = \langle\psi|i\partial_t|\psi\rangle - \langle\psi|H|\psi\rangle - \lambda\langle\psi|\psi\rangle, \tag{5.9}$$

where λ is a Lagrange multiplier which enforces conservation of probability. At time $t = -\infty$, where $\gamma(t) = 0$, we take $g_n = 0$, and choose f_n to minimize $\langle\psi|H|\psi\rangle$,

$$\begin{aligned} \lambda f_n = &-t_a z\left(\sqrt{n}\alpha^* f_{n-1} + \sqrt{n+1}\alpha f_{n+1}\right) \\ &+ \left(\frac{U_{aa}}{2} n(n-1) - \mu n\right) f_n, \end{aligned} \tag{5.10}$$

where

$$\alpha = \sum_n \sqrt{n} f_n^* f_{n-1}. \tag{5.11}$$

Solving the subsequent dynamics to quadratic order in γ, one finds

$$\Gamma(t) = N_s \int dt' \, \gamma(t)\gamma(t')\chi^{(R)}(t-t'), \tag{5.12}$$

where the retarded response function is

$$\chi^{(R)}(t) = \frac{1}{i}\sum_n \sqrt{n}\left(G_n^*(t) f_n - G_n(t) f_n^*\right). \tag{5.13}$$

The Green's functions $G_n(t)$ satisfy the equations of motion for the g_n's in the absence of an RF field, but in the presence of a delta function source, and boundary condition $G_n(t) = 0$ for $t < 0$. The relevant equations are simplest in Fourier space, where $G_n(\omega) = \int dt e^{i\omega t} G_n(t)$ obeys

$$\sqrt{n} f_n = -\omega G_n + \sum_m \Lambda_{nm} G_m \tag{5.14}$$

where $\Lambda = \bar{\Lambda} + \Theta$ is a Hermitian matrix. The tridagonal part $\bar{\Lambda}$ is

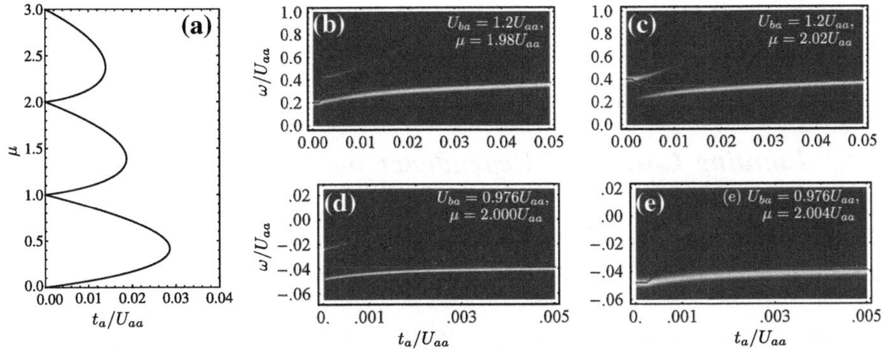

Fig. 5.2 (Color online) Homogeneous system's spectral density as a function of ω/U_{aa} and t_a/U_{aa} (*whiter* indicates larger spectral density) compared with sum rule prediction (*red, single line*). Delta functions are broadened to Lorentzians for visualization purposes. **a** Ground state phase diagram within the Gutzwiller approximation, for reference. We take $U_{ba} = 1.2U_{aa}$ and $t_b = t_a$, with **b** $\mu = 1.98U_{aa}$ and **c** $\mu = 2.02U_{aa}$. We take parameters corresponding to typical ^{87}Rb experiments: $U_{ba} = 0.976U_{aa}$ and $t_b = t_a$, and take **d** $\mu = 2.000U_{aa}$ and **e** $\mu = 2.004U_{aa}$. In both cases, a double peak structure is visible, but the region of the phase diagram in which it is important is much smaller for ^{87}Rb parameters than for part figures (**b**, **c**)'s parameters

$$\bar{\Lambda}_{n,n+1} = -zt_a\alpha\sqrt{n} \tag{5.15}$$

$$\bar{\Lambda}_{n,n-1} = -zt_a\alpha^*\sqrt{n-1} \tag{5.16}$$

$$\begin{aligned}\bar{\Lambda}_{nn} &= -\mu n - \lambda + \frac{U_{aa}}{2}(n-1)(n-2) \\ &\quad + U_{ab}(n-1).\end{aligned} \tag{5.17}$$

The remaining contribution, Θ, is

$$\Theta_{nm} = -zt_b f_{n-1} f^*_{m-1}. \tag{5.18}$$

Specializing to the case where $\gamma(t) = \gamma e^{i\omega t}$, the response is given in terms of normalized eigenvectors v_m, with $\sum_m \Lambda_{nm} v_m^{(j)} = \epsilon_j v_n^{(j)}$. It takes the form of a sum of delta-functions,

$$I(\omega) = \sum_j \left(\sum_m \sqrt{m} f_m v_m^{(j)} \right)^2 \delta(\omega - \epsilon_j). \tag{5.19}$$

The f_n's are found at each point in the phase diagram by starting with a trial α, solving Eq. 5.10, then updating α via Eq. 5.11 and iterating. We find that almost all spectral weight typically lies in only one or two peaks. Figure 5.2 shows sample spectra, whose general behavior is discussed in more detail in Sect. 5.4.2. The superfluid near the Mott state displays a multi-modal spectrum, but in the weakly interacting limit only a single peak is seen. An avoided crossing is clearly visible in these plots. Taking moments of $\chi^R(\omega)$, we see that Eq. 5.5 is satisfied.

5.4.2 Limiting Cases and Dependence on Final State Interactions

5.4.2.1 Bimodality's Evolution Over the Phaes Diagram

Although finding the spectrum in Eq. 5.19 is a trivial numerical task, one can gain further insight by considering limiting cases. First, when $U_{ab} = U_{aa}$ and $t_a = t_b$ the system possesses an $SU(2)$ symmetry. In this limit we find that $G_n(t) = -i\sqrt{n}f_n\theta(t)$ is constant for $t > 0$. Thus our approximation gives a spectrum $I(\omega)$ which is proportional to $\delta(\omega)$. This result coincides with the exact behavior of the system: the operator $X = \sum_j b_j^\dagger a_j$ is a ladder operator, $[H, X] = \delta X$, and can only generate excitations with energy δ (set equal to zero in our calculation). The fact that our approximations correctly capture this behavior is nontrivial: in a field theoretic language one would say that our equation of motion approach includes the vertex corrections necessary for satisfying the relevant "Ward identities" [21–23].

The current ^{87}Rb experiments are slightly perturbed from this limit, with $(U_{ab} - U_{aa})/U_{aa} \approx -0.024$ and $t_b = t_a$. We find that the δ-function is shifted by a frequency proportional to $U_{ab} - U_{aa}$, but that the total spectral weight remains concentrated on that one frequency: the sum of the spectral weights at all other frequencies scale as $\eta \equiv (U_{ab} - U_{aa})/(zt_a)$. Consequently it is an excellent approximation to treat the spectrum as a delta-function, and our RPA calculation reduces to the results in Ref. [5]. We emphasize however that other atoms, such as Cesium, can be in a regime where η is large.

When $\eta \ll 1$ (that is, rapid tunneling rates $t_a \gg (U_{ab} - U_{aa})/z$) the spectral weight is concentrated in a single, shifted peak. Conversely, when $\eta \gg 1$ (that is, slow tunneling rates $t_a \ll (U_{ab} - U_{aa})/z$) the spectral weight is generically multimodal. This behavior can be understood as an analog of "motional narrowing" [24]. Our cartoon Fig. 5.1 gives two excitation processes, and the two states can convert between themselves at a rate $\sim zt_a$. one will only be to separately resolve the two states if the splitting between them is large compared to the rate of the interconversion.

5.4.2.2 Analytic Solution in Deep Lattice Limit

We gain further insight by considering the superfluid near the Mott phase with $t_a/U_a \ll 1$. Here one can truncate the basis to two states with total particle number n and $n+1$ on each site. In this truncation, our calculation is the mean field implementation of the picture in Fig. 5.1. Then the f_n's and G_n's can be found analytically: one only needs to solve 2×2 linear algebra problems. In the $t_b = 0$, $U_{ab} = U_{aa}$ limit, this is similar to Ref. [6]'s approach, but includes the hopping self consistently, allowing us to satisfy the sum rule Eq. 5.5. This truncation is exact in the small t_a limit, and yields

$$\chi^{(R)}(\omega) = A_+\delta(\omega - \omega_+) + A_-\delta(\omega - \omega_-) \tag{5.20}$$

with

$$\omega_\pm = \frac{\epsilon_1 + \epsilon_2}{2} \pm \sqrt{\Delta^2 + \left(\frac{\epsilon_1 - \epsilon_2}{2}\right)^2} \tag{5.21}$$

where

$$\begin{aligned}
\epsilon_1 &\equiv (U_{ab} - U_{aa})(n-1) + zt_a f_{n+1}^2(n+1) \\
\epsilon_2 &\equiv (U_{ab} - U_{aa})n + z[t_a(n+1) - t_b]f_n^2 \\
\Delta &\equiv -\sqrt{n(n+1)}t_a z f_n f_{n+1}.
\end{aligned} \tag{5.22}$$

if $n \geq 1$ and

$$\begin{aligned}
\epsilon_1 &\equiv zt_a f_1^2 \\
\epsilon_2 &\equiv z(t_a - t_b)f_0^2 \\
\Delta &\equiv 0.
\end{aligned} \tag{5.23}$$

if $n = 0$ (here, only the ϵ_2 peak has non-zero spectral weight). We omit the cumbersome analytic expressions for the spectral weights $A_\pm$. The spectrum consist of two peaks—hybridized versions of the excitations caricatured in Fig. 5.1. One can identify ϵ_1 and ϵ_2 as the energies of those caricature processes, recognizing that the hybridization term, Δ, grows with t_a. The avoided crossing between these modes is evident in Fig. 5.2.

5.4.3 Inhomogeneous Spectrum

We model the trapped spectrum through a local density approximation. We assume that a given point in the trap has the properties of a homogeneous gas with chemical potential $\mu(r) = \mu_0 - V(r)$. In Fig. 5.3 we show the density profile and the spectrum corresponding to each point in space. Also shown is the trap averaged spectrum.

When U_{ab} is slightly smaller than U_{aa}, corresponding to ^{87}Rb and illustrated in Fig. 5.3e, the spectrum is unimodal at each point in space. The shift, given by the sum rule in (5), maps out the density seen in Fig. 5.3a. For larger $|U_{ab} - U_{aa}|$, illustrated in Fig. 5.3b, the spectrum in the superfluid regions is distinctly bimodal. If, as in Fig. 5.3c, one makes $U_{ab} = U_{aa}$, but sets $t_b \neq t_a$, the spectrum reflects the

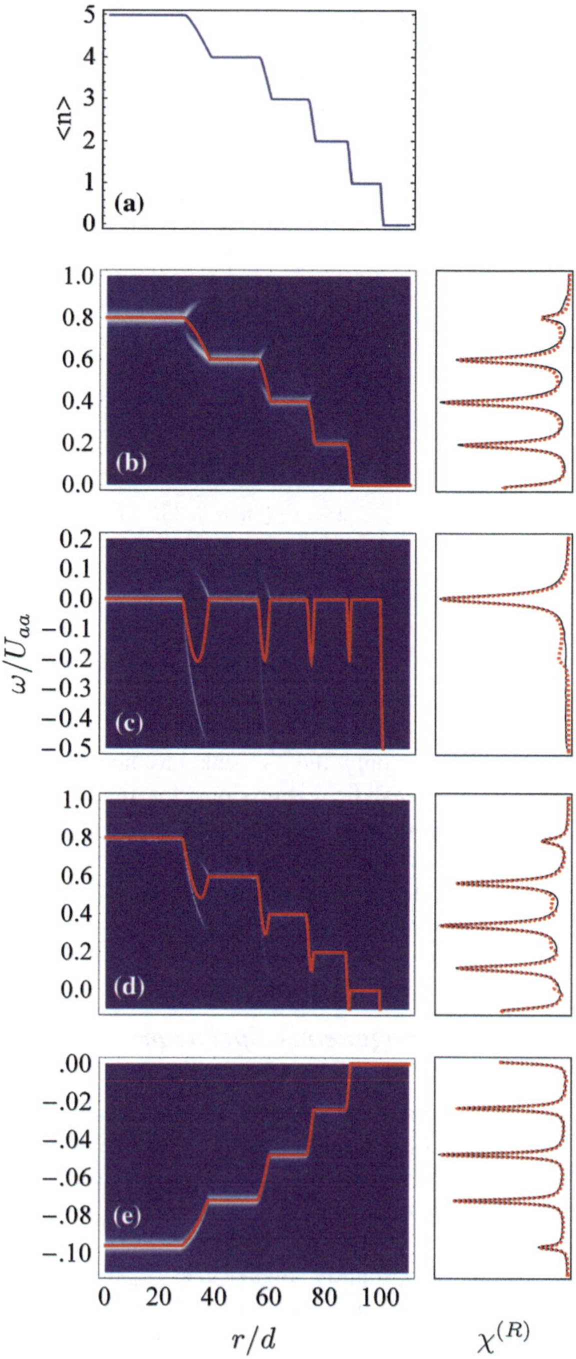

Fig. 5.3 (Color online) **a** Density n as a function of distance to trap center rescaled by the lattice spacing, r/d, in a local density approximation. For all subfigures, we take $t_a/U_{aa} = 0.004$, which is moderately smaller than the tip of the first Mott lobe. **b–e** *Left*: spectrum of a homogeneous gas with density $n(r)$, representing the spatially resolved spectrum observed in an experiment on a trapped gas. Horizontal axis is position, vertical is frequency, color from dark to light represents increasing spectral density. *Continuous* (*red*) *curve* denotes sum rule result for $\langle\omega\rangle$. We round the δ-functions to Lorentzians for visualization. *Right*: trap-averaged spectrum for a 3D trap within our RPA (*black, solid line*) compared with sum rule (*red, dashed line*). **b** $U_{ab} = 1.2U_{aa}, t_b = t_a$; **c** $U_{ab} = U_{aa}, t_b = t_a + 0.1U_{aa}$; **d** $U_{ab} = 1.2U_{aa}, t_b = t_a + 0.1U_{aa}$; **e** ^{87}Rb parameters: $U_{ab} = 0.976U_{aa}, t_b = t_a$

condensate density rather than the particle density. One only sees spectral shifts in the superfluid regions—finding spectral weight gradually shift from one peak to another as one moves in space. In Fig. 5.3d we illustrate what happens when both $U_{ab} \neq U_{aa}$ and $t_b \neq t_a$. One sees a pattern which is roughly a superposition of Fig. 5.3b, c.

In all cases, the bimodality of the homogeneous spectrum is quite effectively washed out by the inhomogeneous broadening of the trap. On the other hand, if one spatially images the atoms flipped into the b state as in Ref. [3], there is a clear qualitative signature of the bimodality. If one excites the system with an RF pulse whose frequency lies between the resonant frequencies of two Mott plateaus, one will excite two "shells" of atoms. These shells should be clearly visible, even in column integrated data.

5.5 Conclusions and Discussion

In this paper we have shown that the RF spectra of a homogeneous Bose gas in an optical lattice will have two (or more) peaks in the superfluid state when the parameters are tuned close to the superfluid-Mott insulator phase transition. Physically, this bimodality is a result of the strong correlations in the system. These correlations result in two distinct forms of excitations (which are strongly hybridized): those involving "core" atoms, and those involving delocalized atoms. When $\eta = (U_{ab} - U_{aa})/(zt_{aa})$ is small, such as in typical ^{87}Rb experiments, this bimodality is absent.

Our approach, based upon applying linear response to a time dependent Gutzwiller mean field theory, is both simple and quite general. It allows arbitrary interactions between both spin states, and it allows arbitrary spin-dependent hopping rates. The major weakness of the theory is that it fails to fully account for short range-correlations: the atoms are in a quantum superposition of being completely delocalized, and being confined to a single site. The physical significance of this approximation is most clearly seen when one considers the case where the final-state atoms have no interactions, $U_{ab} = 0$, and see no trap or lattice. Imaging the b-atoms after a time-of-flight is analogous to momentum resolved photoemission [25], and would reveal the dispersion relationship of the single-particle excitations. The fact that the spectrum consists of two sharp peaks means that all of the non-condensed atoms are approximated to have the same energy. One will also see that their momentum is uniformly distributed throughout the first Brillioun zone. In the strong lattice limit, where the bandwidth is small, this approximation is not severe.

Acknowledgements We thank Sourish Basu, Stefan Baur, Stefan Natu, Eliot Kapit, Wolfgang Ketterle, Kuei Sun, Smitha Vishveshwara, Henk Stoof, Ian Spielman, and Mukund Vengalattore for useful discussions. This material is based upon work supported by the National Science Foundation through grant No. PHY-0758104, and partially performed at the Aspen Center for Physics.

References

1. D. Jaksch, C. Bruder, J.I. Cirac, C.W. Gardiner, P. Zoller, Phys. Rev. Lett. **81**, 3108 (1998)
2. M.P.A. Fisher, P.B. Weichman, G. Grinstein, D.S. Fisher, Phys. Rev. B **40**, 546–570 (1989)
3. G.K. Campbell, J. Mun, M. Boyd, P. Medley, A.E. Leanhardt, L.G. Marcassa, D.E. Pritchard, W. Ketterle, Science **313**, 649 (2006)
4. I. Bloch, Nat. Phys. **1**, 23 (2005)
5. K.R.A. Hazzard, E.J. Mueller, Phys. Rev. A **76**, 063612 (2007)
6. K. Sun, C. Lannert, S. Vishveshwara, Phys. Rev. A **79**, 043422 (2009)
7. D.G. Fried, T.C. Killian, L. Willmann, D. Landhuis, S.C. Moss, D. Kleppner, T.J. Greytak, Phys. Rev. Lett. **81**, 3811 (1998)
8. Y. Ohashi, M. Kitaura, H. Matsumoto, Phys. Rev. A **73**, 033617 (2006)
9. M.Ö. Oktel, L.S. Levitov, Phys. Rev. Lett. **83**, 69 (1999)
10. I.H. Deutsch, P.S. Jessen, Phys. Rev. A **57**, 1972–1986 (1998)
11. L. Pollet, S. Rombouts, K. Heyde, J. Dukelsky, Phys. Rev. A **69**, 043601 (2004)
12. S. Bergkvist, P. Henelius, A. Rosengren, Phys. Rev. A **70**, 053601 (2004)
13. S. Wessel, F. Alet, M. Troyer, G.G. Batrouni, Phys. Rev. A **70**, 053615 (2004)
14. G.G. Batrouni, V. Rousseau, R.T. Scalettar, M. Rigol, A. Muramatsu, P.J.H. Denteneer, M. Troyer, Phys. Rev. Lett. **89**, 117203 (2002)
15. B. DeMarco, C. Lannert, S. Vishveshwara, T. Wei, Phys. Rev. A **71**, 063601 (2005)
16. N. Dupuis, K. Sengupta, Phys. B Condens. Matter **404**, 517 (2009)
17. K. Sengupta, N. Dupuis, Phys. Rev. A **71**, 033629 (2005)
18. S. Konabe, T. Nikuni, M. Nakamura, Phys. Rev. A **73**, 033621 (2006)
19. C. Menotti, N. Trivedi, Phys. Rev. B **77**, 235120 (2008)
20. M. Öktel, T.C. Killian, D. Kleppner, L.S. Levitov, Phys. Rev. A **65**, 033617 (2002)
21. C.J. Pethick, H.T.C. Stoof, Phys. Rev. A **64**, 013618 (2001)
22. G. Baym, Phys. Rev. **127**, 1391 (1962)
23. J. Zinn-Justin, *Quantum Field Theory and Critical Phenomena*, 4th edn. (Oxford University Press, New York, 2002)
24. C.P. Slichter, *Principles of Magnetic Resonance* (Springer, New York, 1990)
25. J.T. Stewart, J.P. Gaebler, D.S. Jin, Nature **454**, 744 (2008)

Chapter 6
Radio-Frequency Spectra at Finite Temperature, Fluctuation-Response Relations, and Proposed Applications

6.1 Chapter Overview

In this chapter, I further develop the theory necessary to apply RF spectra techniques to overcoming experimental challenges in ultracold atomic systems.

Firstly, I address the most salient features of the finite temperature theory of bosons in optical lattices through the Mott insulator-superfluid phase transition. Here I focus on the Mott insulator and superfluid regimes; the quantum critical regime is addressed in Chap. 13. The results suggest applications, including thermometry and perhaps cooling.

Secondly, I develop ideas to relate sum rules for RF spectra in a trap to local susceptibilities, density profiles, and spatial correlations of the bosonic lattice system. I give both exact relations of the spectra to high order correlation functions, as well as simple approximations relating the spectra to simpler quantities.

Finally, I show how using longer range interactions, one can probe arbitrary n-point density correlations of multiple sites. Although the experimental difficulty to resolve these increases with increasing n and distance spanned in the correlation function, realistic experimental parameters indicate that it may usefully probe $n \lesssim 5$ correlators up to a 3–4 sites apart.

6.2 Introduction

To experimentally examine any ultracold atomic state of matter presents two challenges: creating it and measuring its properties. For one, achieving sufficiently low enough temperatures is difficult [1]. To know the system temperature is also difficult, and single-digit accuracy is state-of-the art. Finally, to probe a system is to measure its correlation functions, and while the cold atom systems lend themselves to measuring real- and k-space density distributions, many other

K. R. A. Hazzard, *Quantum Phase Transitions in Cold Atoms and Low Temperature Solids*, Springer Theses, DOI: 10.1007/978-1-4419-8179-0_6,

correlation functions necessary to diagnose phase transitions and properties of phases are presently inaccessible. As notable examples, consider the momentum dependence of the anomalous 1-particle Green's function in the fermion Hubbard model (i.e., is there d-wave superconductivity?), and n-particle Green's functions with $n > 2$ for general liquids (interesting for example because many approximations amount to assuming that the $n > 2$ Green's functions have a certain structure, even though this structure is not well understood due to the experimental inaccessibility.)

I suggest multiple applications and develop the necessary theory for improving our ability to characterize these correlation functions ultracold atoms. Specifically, I show how RF spectra are related to local compressibilities and various spatial correlations, and how this may be used for thermometry and cooling.

In the process, I provide a theoretical toolbox to enable future experiments: Sect. 4.3 outlines the necessary background for calculating RF spectra in this context. Sect. 6.4 treats finite temperatures, and Sect. 6.5 suggests applications.

6.3 RF Spectra Introduction

Hamiltonian. The Bose-Hubbard (BH) model describing lattice bosons with two internal states is [2]

$$\begin{aligned} H = -t_a \sum_{\langle i,j \rangle} a_i^\dagger a_j + \sum_i \left[\frac{U_a}{2} n_{i,a}(n_{i,a} - 1) + V_{i,a} n_{i,a} \right] \\ - t_b \sum_{\langle i,j \rangle} b_i^\dagger b_j + \sum_i \left[\frac{U_b}{2} n_{i,b}(n_{i,b} - 1) + V_{i,b} n_{i,b} \right] \\ + U_{ab} \sum_i n_{i,a} n_{i,b} + \text{H.c.} \end{aligned} \tag{6.1}$$

where $n_{i,\alpha} \overset{\text{def.}}{\equiv} \alpha_i^\dagger \alpha_i$, and the a's and b's are boson operators. The t_α's are hopping rates of species α, and $U_{\alpha\beta}$ the interaction of species α and β. I define $V_{i,\alpha} = \mathcal{V}_{i,\alpha} - \mu_\alpha$ where $\mathcal{V}_{i,\alpha}$ is the external potential at site i for species α, so H is the $T = 0$ grand canonical free energy.

Approximations. In RF spectroscopy, the atoms are initially all in the a-internal state, and then a pulse drives some into the b-state. Thus, the spectrum depends on the state of the all-a system and parameters appearing explicitly in Eq. 4.1. As in previous chapters, I use the Gutzwiller mean-field (GMF) theory and local density approximation (LDA). I briefly review the these approximations in notation suitable for the current discussion. The mean-field formulation of GMF assumes the fluctuations of the operators from their expectation values couple negligibly in the Hamiltonian to neighboring sites. Then one can write

$a_i^\dagger = \langle a_i^\dagger \rangle + \left(a_i^\dagger - \langle a_i^\dagger \rangle \right) \approx \langle a_i^\dagger \rangle$ for the operators "at other sites," so that the mean-field Hamiltonian is, assuming unbroken translational symmetry,

$$H_{\mathrm{MF}} = \sum_i \left[-qta_i^\dagger \langle a \rangle + U \frac{n_i^2}{2} + V_i n_i + \mathrm{H.c.} \right] \tag{6.2}$$

where q is the number of nearest neighbors.

The LDA is used to calculate the spatial dependence of thermodynamic quantities, especially the density. That is, we take properties P(local density, energy, etc.) at a spatial location $\mathbf{r}$, to be $P(\mathbf{r}) = P(\mu - V(\mathbf{r}))$ where $P(\mu)$ is the observable as a function of the chemical potential in the homogeneous system with no external potential and $V(\mathbf{r})$ is the external potential.

RF spectra background. The *average* frequency shift Δ is a shift in the energy between a and b states due to the inter-particle interactions. As discussed in Chap. 4, the general formula for Δ in a homogeneous system is [3, 4]

$$\Delta = (U_{ab} - U_a) \frac{\langle n(n-1) \rangle}{\langle n \rangle}.$$

where n is the site occupation. In the SF, MI, and normal fluid (NF) states, it is [5, 6]

$$\Delta_{\mathrm{SF}} = (U_{ab} - U_a)\langle n \rangle \tag{6.3}$$

$$\Delta_{\mathrm{MI}} = (U_{ab} - U_a)(\langle n \rangle - 1) \tag{6.4}$$

$$\Delta_{\mathrm{NF}} = 2(U_{ab} - U_a)\langle n \rangle. \tag{6.5}$$

In a trap, the system will take on a variety of energy shifts, corresponding to different points in space. Shining a laser with frequency near the a to b transition energy and counting the number of b states yields the number of atoms with mean energy shift $\Delta = \omega$. I denote this trap averaged RF-spectrum $N_b(\omega)$. To calculate $N_b(\omega)$, we discretize in r and calculate $\Delta(r)$ and $n(r)$ numerically within GMF+LDA. Then for each energy range $\Delta(r)$ we find the corresponding radial shells (of width δr) and sum up $n(r)4\pi r^2 \delta r$ over these shells to obtain $N_b(\omega)$.

6.4 Finite Temperature Superfluid

This section calculates RF spectra for the weakly interacting superfluid at finite temperature. Section 6.5 treats the finite temperature MI state.

Neglecting the "hybridization" of the states that occur after transfer of atoms, the NF and the condensate give independent contributions to the spectrum. Oktel and Levitov have treated the hybridization [7]. In general, one needs Oktel and Levitov's full theory, but in many regimes or for a qualitative understanding, it suffices to neglect this hybridization.

In the dilute superfluid, finite temperature leads to a thermal gas of excitations coexisting with a condensate, whose density is modified from zero temperature. We calculate these densities using a semiclassical version of Hartree-Fock (HF) mean-field theory.[1] In this limit, and neglecting excitation hybridization, the condensate and excitation densities determine RF spectra since Eqs. 4.5 and 6.5 relate the condensate density and the excitation density to the location of the RF spectral peaks. As an example, a *homogeneous* SF's RF spectrum has two peaks: one due to the condensate and one (at twice the excitation frequency) due to the NF.

The semiclassical HF excitation density is is [8, 9]

$$n_{\rm ex}(\mathbf{r}) = \int \frac{d\mathbf{p}}{(2\pi\hbar)^3} \frac{1}{e^{(\epsilon_{\mathbf{p}}(\mathbf{r})-\mu)/k_BT} - 1} \tag{6.6}$$

where the integral is over all momenta and the excitation energy is (more on this in a moment) $\epsilon_{\mathbf{p}}(\mathbf{r}) = p^2/2m + 2n(\mathbf{r})U_a + V(\mathbf{r})$ with the total density n being the sum of the condensate and excitation densities, $n = n_0 + n_{\rm ex}$. Finally, the condensate density is

$$n_0(\mathbf{r}) = \frac{\mu - V(\mathbf{r})}{U_\alpha} - 2n_{\rm ex}(\mathbf{r}).$$

The chemical potential is determined to reproduce the system's total atom number $N = \int d\mathbf{r}[n_0(\mathbf{r}) + n_{\rm ex}(\mathbf{r})]$. Then all of these equations are solved self-consistently. In words, this calculation is easily summarized: the condensate density is given by the Thomas-Fermi approximation, as is the density of the thermal excitations; in each case, interactions are treated within the HF approximation.

I briefly outline the approximations going into this calculation. The kinetic part of the single particle energy $\epsilon_{\mathbf{p}}$ is taken to have the "free" particle form, $p^2/(2m)$, which is justified if (1) the temperatures are small compared to the bandwidth so that the dispersion may be expanded to quadratic order (the square lattice assures that the dispersions is isotropic to quadratic order) and (2) the temperatures are large enough that the interactions are in the single-particle rather than Bogoliubov regime—the criterion for this is that $T \gg ms^2 \simeq nU_\alpha$ where s is the sound speed. The interaction term with the total density takes the form $2n(\mathbf{r})U_a$ in the HF approximation, which is also qualitatively accurate over a wide temperature range. [9] As long as the thermal de Broglie wavelength and coherence length are small compared to the trap's characteristic length scales, the Thomas-Fermi approximation is accurate for sufficiently large particle number [8].

---

[1] Technically, since the scattering length is renormalized, it is something more akin to Brueckner-Hartree-Fock.

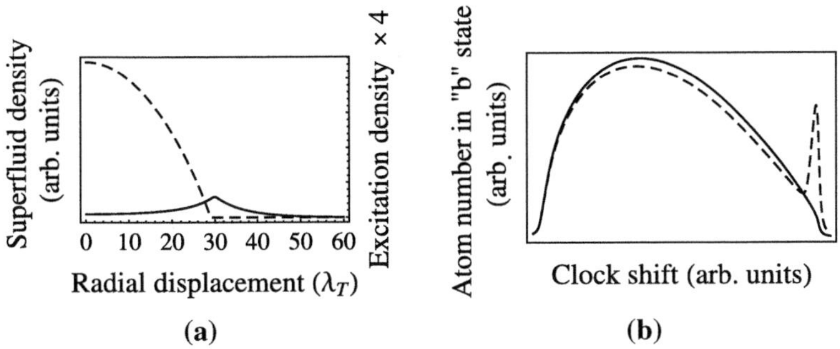

Fig. 6.1 **a** Condensate and excitation densities of a finite temperature superfluid at a temperature of 30 in units of the coupling constant U_α, the radial displacement is in units of the thermal de Broglie wavelength λ_T, and the trapping potential is αr^2 with $\alpha = 10^{-2}U_\alpha/\lambda_T^2$. The excitation density is rescaled to make it more visible. **b** Corresponding RF spectra at temperatures $T = 1, 30$ for the solid and dashed curves respectively. **b** The peak at low densities comes from the normal fluid tail

Figure 6.1 shows a characteristic example of a finite temperature SF RF spectrum, as well as the condensate and excitation density profiles. The most salient feature that emerges is the presence of a peak associated with nearly zero density—this occurs due to the tail of thermally populated NF at the edge of the trap.

I conclude the analysis of the finite temperature SF spectra with two observations: (1) the NF peak may improve thermometry of weakly interacting gases; by measuring the peak height relative to the SF spectral background, one might accurately measure the temperature. (2) Going beyond the HF theory, one finds a finite thermal excitation density at the center of the trap at low enough temperatures that then peaks at the edge of the condensate cloud, and then decreases beyond the condensate (see, e.g., Ref. [9]). Thus, a temperature-dependent *kink* appears in the RF spectrum at an energy corresponding to the excitation density at the center of the trap. Since this kink persists, and its magnitude changes, all the way to zero temperature, this may enable very-low temperature thermometry. Furthermore, it provides experimental access to the beyond-HF complexities of the finite-temperature Bose-liquid that control the density at the center.

6.5 Applications

6.5.1 Thermometry with the Mott Insulator

6.5.1.1 Finite Temperature Mott Insulator RF Spectrum

Accurate thermometry is a crucial area for development in ultracold atoms: their use as quantum simulators for many-body physics requires an accurate knowledge and control of the temperature. We propose a sensitive, low temperature thermometry method based on the deep MI state's RF spectrum.

We first review the elementary theory of the finite-temperature MI as $t/U \to 0$. When t = 0 each lattice site decouples from the others, and we can solve the single site problem to obtain the excitation spectrum. The eigenstates are then Fock states, $|n\rangle = (1/\sqrt{n!})(a^\dagger)^n|0\rangle$ where $|0\rangle$ is the vacuum, with energies

$$E_n = \frac{U_a}{2} n(n-1). \tag{6.7}$$

Applying the sum rule directly for the $t/U \ll 1$ case, the system is "deep in the Mott state" and the average spectral shift frequency is just $(U_a - U_{ab})(\langle n \rangle - 1)$—the formula argued previously, but with the expectation including a thermal as well as quantum expectation [3, 4]. But it is not only the center of the RF spectrum we are interested in—we will see that extra structure develops in the form of peaks at the various integer densities, even in a homogeneous system.

For a state initially in a incoherent combination of eigenstates $|n\rangle$ with probabilities P_n, the result is that one can associate a spectral shift Δ_n with each initial eigenstate $|n\rangle$, with only a quantum average—no thermal average. Then the resulting RF spectrum consists of the sum over n of peaks at energies Δ_n with height weighted by P_n.

Thus to determine the RF spectrum we need to calculate P_n. In thermal equilibrium, the probabilities P_n are

$$P_n = \frac{e^{-\beta(E_n - \mu n)}}{\sum_{n=0}^{\infty} e^{-\beta(E_n - \mu n)}}.$$

Rewriting $\beta(E_n - \mu n) = \frac{\beta U_a}{2}\left[(n-n_0)^2 - n_0^2\right]$ with $n_0 = \mu/U_A - 1/2$ we see that the homogeneous system's RF spectrum is a set of peaks associated with integer densities enveloped by a linear function times a Gaussian with mean n_0 and standard deviation $\sigma \overset{\text{def.}}{\equiv} (\beta U_a)^{-1/2}$ (the normalization is unimportant):

$$\mathcal{I}_{\text{hom}}(\mu) = \mathcal{Z}(\mu)\, n e^{-(n-n_0(\mu))^2/2\sigma^2} \tag{6.8}$$

where $\mathcal{Z}(\mu) = \sum_{n=0}^{\infty} e^{-(n-n_0)^2/2\sigma^2}$.

Figure 6.2 shows a characteristic finite temperature RF spectrum for a homogeneous system with a chemical potential $\mu = 1.8U$, corresponding to site filling n = 2 at zero temperature. The delta function peaks are convolved with Lorentzians to simulate line broadening due to the finite time duration (bandwidth) of the probe pulse or higher order tunneling effects.

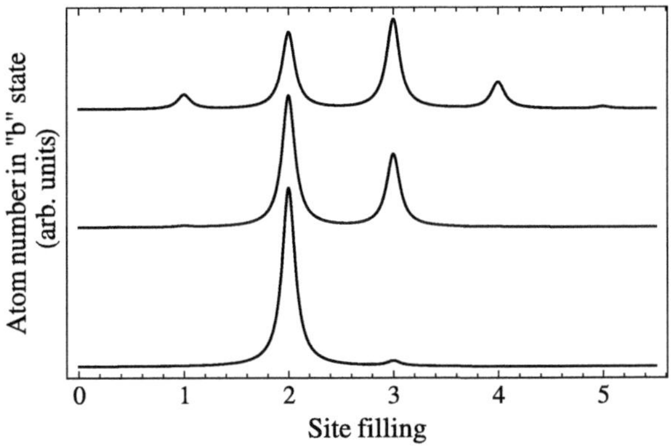

Fig. 6.2 Finite temperature RF spectrain the deep Mott limit for a *homogeneous system*, at temperatures $T = 0.05, 0.2, 0.8$ in units of U_a, from *bottom* to *top*. The appearance of multiple peaks signals the occupation of states with different site fillings. The chemical potential is $\mu = 1.8$ (corresponding to a $n = 2$ site filling at zero temperature)

6.5.1.2 Finite Temperature Mott RF Spectra in a Trap

To calculate RF spectra in a harmonic trapping potential, we use an LDA approximation: calculate the homogeneous RF spectrum for local chemical potential $\mu_{\text{eff}}(\mathbf{r}) = \mu - \alpha r^2$ for each position and add these. Hence, the RF spectrum in the trapped system is

$$\mathcal{I}_{\text{trap}} \propto \int_0^\infty 4\pi r^2 dr\, \mathcal{I}_{\text{hom}}(\mu_{\text{eff}}(r)).$$

We numerically integrate $\mathcal{I}_{\text{hom}}$ from Eq. 6.8.

To apply this to typical experiments, we must fix particle number N rather than μ as the temperature is changed. This is easily accomplished by computing N as a function of the μ and numerically inverting. Actually, in many the temperature and particle number ranges of interest (say $T < 3U_a$, $N > 10^5$), μ is nearly temperature-independent, so it is unnecessary account for this variation to produce quite accurate RF spectra.

Figure 6.3 shows the RF spectra for the trapped MI for various temperatures, at fixed particle number. Figure 6.3(a) gives another warning: although the system is everywhere in the singly- and doubly-occupied MI states, fillings as high as n $= 5$ can appear at temperatures on the order of U_a due to thermal excitations. Hence, the RF spectra appear as if there were $n = 3, 4, 5$ MI shells even though no such shells exist, further complicating the interpretation of the experiments of Ref. [10].

6.5.1.3 Thermometry

As detailed in the introduction, thermometry of ultracold atomic systems is a key goal for the field. Because the MI RF spectrum depends strongly on temperature (e.g., new peaks appear rapidly at some temperature on the order of U_a), it

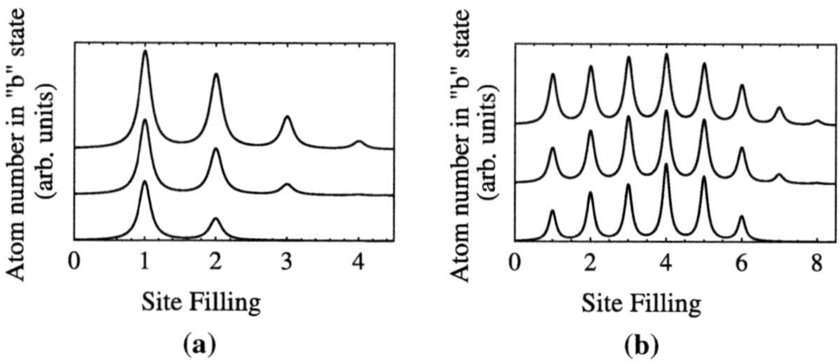

Fig. 6.3 Finite temperature RF spectra in the deep Mott limit for a harmonically trapped system. The temperatures are $T = 0.01, 0.7, 1.5$ in units of U_a from bottom to top. Vertical offsets are for clarity only. The chemical potential is $\mu = 1.4, 5.7$ (corresponding to $n = 2, 6$ zero-temperature site fillings at the trap center) in (**a**) and (**b**) respectively

suggests looking into this system as a thermometer. I argue that it has many other desirable features for thermometry.

Review and drawbacks of existing methods. Before describing the proposed thermometry method, we first review the typical method for doing thermometry—measuring density in the tails of the cloud—and highlight its drawbacks. The system becomes dilute in this region, and hence the density just decays, in the LDA approximation, as an exponential of the local chemical potential. This leads to Gaussian decay as a function of position in a harmonic trap. By fitting the Gaussian's length scale ℓ, one obtains the temperature.

There are several drawbacks to this method, however:

1. Only a small fraction of atoms at the surface of the cloud are imaged, leading to small signal-to-noise.
2. One must inverse Abel transform columnar integrated densities to obtain the true density profile. This leads not only to additional analysis, but more importantly to further propagation of error, and sensitivity to systematic errors from anisotropy—the Abel transform assumes isotropy about at least one axis. One can also attempt to work directly fitting the column integrated density, but this is even less sensitive to temperature than the density.
3. The trap is most anharmonic in the tails; only densities inside some radius r_o feel a harmonic potential. Let r_i be the distance beyond which the density is dilute enough to be Gaussian. To accurately measure temperature, one must have $\ell/(r_o - r_i) \ll 1$, putting an upper limit on ℓ's measurable and thus a lower limit on temperatures measurable.
4. If ℓ is smaller than the laser's spatial resolution ($\sim 10\ \mu$m), the length scale/temperature is unresolvable. Since $\ell \sim 1/T$, this imposes a fundamental limit on the lowest measurable temperatures. For a trap with frequency $\omega_0 \sim 100\ \text{s}^{-1}$ and Rubidium atoms, one find that the lowest resolvable temperatures are about 10 nK. *Much* lower temperatures are desired for simulating strongly correlated

and exotic spin systems. State-of-the-art single site imaging techniques on lattices optimistically might improve this to 0.1 nK.

5. The edges of the cloud are often the furthest from the region of interest, and thus will be slowest to equilibrate with it.
6. The measurement is not robust: (a) One needs to know the trapping frequency ω_0, as errors in ω_0 propagate to errors in temperature. (b) One needs a globally harmonic trap, and anharmonicities are not automatically detected if the trap is harmonic in the tail region (as any sufficiently small region is); if the system looks harmonic locally but is anharmonic globally then the density locally looks Gaussian, but the appropriate frequency is the *local frequency* of the anharmonic trap, and the temperature will be correspondingly wrong if one uses the frequency calibrated from the center of the trap. (c) There is no way to tell from the data whether the trap is isotropic over the two directions assumed to be identical.

Thermometry proposal. I will propose equilibrating a MI with a system to be measured and using the MI's RF spectrum as a thermometer. I will show that this thermometer is less restricted by each difficulty than measuring wings of the cloud. First, one equilibrate the MI with the measured system; by tuning the interaction strength through a Feshbach resonance or other means, the thermometer can thermally equilibrate with the system to be probed, but then turned off so that the dynamics are not affected after the measured system. One then measures the MI's RF spectrum and compares with the spectrum calculated in the previous section to measure the temperature. Because of the strong qualitative change in the RF spectrum with temperature, one can fit reliably.

Perhaps the optimal situation for fitting is where the interaction energy is large enough so that at zero temperature there are no doubly-occupied states, but plenty of singly-occupied states. Then by increasing the temperature, a qualitative feature appears: a new peak at $|n = 2\rangle$, which grows with increasing temperature. The relative peak heights allows one to infer the temperature.

We enumerate ways in which this technique circumvents the problems of the density tail-measurements:

1. In RF spectra measurements, all atoms participate rather than just the tails, and consequently one achieves better signal-to-noise.
2. Features survive the columnar integration (and the rest of the trap integration). Moreover, anisotropy presents no problems so long as the trap is harmonic. The calculations already given still give correct RF spectrum, because the surface area of an ellipsoid is still proportional to r^2 (with r the distance to the center of the trap in some preferred direction); it's just the irrelevant numerical prefactor that's modified.
3. There are important contributions from the entire trap, so anharmonic tails are fairly unimportant, especially at low temperatures.
4. There is no intrinsic low-temperature limit to the technique. The lower limit is simply set by the value of U_a. The goal is to achieve the minimum U_a, since the

finite temperature effects are significant only when T is comparable or larger than U_a, while remaining in the MI limit ($t \ll U_a$). Hence, one needs t very small compared to U_a even when U_a is small. This can be effected by increasing the lattice spacing—this suppresses U_a as well as t/U.
5. RF spectra measure over the entire trap, so if some regions with few particles (e.g. the tails) are out of equilibrium, it causes less error in the measurement.
6. The method is robust: problems are easily diagnosed. By changing the optical lattice depth U_a changes in a predictable way and hence one can calculate the expected changes in the RF spectra, assuming a harmonic trap. This offers a diagnostic: if the diagrams don't change in the expected manner, then it signals that one has an anisotropic trap, a wrong fit to the theory, or some other problem.

Because of these numerous advantages, particularly the first and the last, we expect this method to provide a substantially more robust and accurate thermometry than measuring the tails of density distributions.

6.5.2 Number Correlations and Local Compressibility

Relating RF spectra to static response functions and correlations provides an alternative, intuitive picture of them. This also leads naturally to the direct measurement of quantities which may otherwise be inaccessible (e.g. a *local* compressibility in a low imaging resolution experiment). The compressibility is

$$\kappa \overset{\text{def.}}{\equiv} \frac{\partial n}{\partial \mu} = \kappa^{-1} \propto \partial^2 \epsilon / \partial n^2$$

with ϵ the energy. Systems (at zero temperature) with a gap to charge excitations are incompressible ($\kappa = 0$). Consequently, the MI is incompressible.

6.5.2.1 RF Spectra-Compressibility Relation (Approximate)

To relate the RF spectrum to the compressibility (as opposed to more complicated response functions)requires an approximation. It assumes that RF spectra are histograms of the number of particles with a local particle density n up to a scale factor. Both the deep MI and SF limits satisfy this exactly, and everywhere else satisfies this approximately. [5, 6] We also assume that a sum rule approach accurately describes the local spectra.

The frequency shift is assumed in this approximation to be linearly related to the density. Thus, I will thus calculate the RF spectrum as a function of density, $\mathcal{I}(n)$, which is then related to $\mathcal{I}(\omega)$ simply by a rescaling of the function's argument. I introduce the cumulative distribution

$$C(\rho) = -\int_{\mathcal{V}_\rho} d\mathbf{r}\, n(\mathbf{r}) \tag{6.9}$$

where $\mathcal{V}_\rho$ is the volume of atoms with density ρ or higher. Thus, the total RF spectral weight for densities in $(\rho, \rho + \delta\rho)$, given by $\mathcal{I}(\rho)\delta\rho$, is proportional to the number of atoms in this density range, $-C'(\rho)\delta\rho$ (the minus sign occurs because $C(\rho)$ decreases as ρ increases) and

$$\mathcal{I}(\rho) = -\frac{dC}{d\rho}.$$

Actually, $I(\rho)$ more properly is $\mathcal{I}(\rho + f)$, since we only assume that the spectral energy shift are proportional to the filling up to a constant offset f, but we will ignore this constant offset in the spectra. In order to relate spectra as a function of ρ to spectra as a function of ω, we need another means of determining the offset, perhaps from theoretical considerations.

Thus the spectrum is

$$\begin{aligned}\mathcal{I}(\rho) &= -\frac{d}{d\rho}\left[\int_0^{r(\rho)} 4\pi r^2 dr\, n(r)\right] \\ &= -4\pi\frac{dr}{d\rho}[r(\rho)]^2\rho \\ &= -4\pi\frac{dr}{d\mu_{\text{eff}}}\frac{d\mu_{\text{eff}}}{d\rho}[r(\rho)]^2\rho \\ &= -4\pi\frac{dr}{d\mu_{\text{eff}}}[\kappa(\rho)]^{-1}[r(\rho)]^2\rho. \end{aligned} \tag{6.10}$$

The derivative is $\left.\frac{dr}{d\mu_{\text{eff}}}\right|_{\mu_{\text{eff}}(\rho)} = -\frac{1}{m\omega^2 r(\rho)}$ so

$$\mathcal{I}(\rho) = \frac{4\pi}{m\omega^2}\frac{\rho r(\rho)}{\kappa(\rho)}. \tag{6.11}$$

This relates RF spectral weight at frequencies corresponding to density ρ and the compressibility of the *homogeneous* system with density ρ. This provides an intuitive way of thinking about RF spectral weight: the RF spectral weight comes from relatively incompressible regions, in which the density varies slowly. As usual, it also holds in an anisotropic trap because the surface area of a constant-density shell still scales as r^2.

Inverting Eq. 6.11 relates the gives κ in terms of the RF spectrum:

$$\kappa(\rho) = \gamma\frac{\rho r(\rho)}{\mathcal{I}(\rho)}$$

where γ is independent of ρ. One can determine $r(\rho)$ by spatially measuring where the RF spectra absorption at energy corresponding to ρ occurs. Hence RF spectra measure κ of the *homogeneous* phase with density ρ. This requires some spatial resolution to determine $r(\rho)$, but is likely significantly less demanding than alternative methods to obtain the compressibility, for example by differentiating density profiles.

This yields a significant measurement tool. Although one can presently measure the *global* compressibility of the trapped system by changing the trap frequency and seeing how the filling distribution changes, the *local* compressibilities are harder to measure. The deep MI system provides a particularly striking example of the limitations of measuring only global compressibility. Even though the MI is everywhere incompressible, with $\kappa(\rho) = 0$, the global compressibility is finite [11]. This is due to the regions at the edges of MI shells where the gap closes; locally, there the system is compressible. Although these regions strictly occupy zero volume, they provide a mechanism for density adjustment by particles moving across these regions, leading to a globally compressible system.

Finally, it is worth relating the compressibility to number fluctuations, giving a tool to measure these or, conversely, another way to think about RF spectra. The site-i compressibility is

$$\begin{aligned}\kappa_i &= \frac{\partial\langle n_i\rangle}{\partial\mu} \\ &= \beta\sum_j\left(\langle n_i n_j\rangle - \langle n_i\rangle\langle n_j\rangle\right)\end{aligned}$$

via usual correlation-response relations.

6.5.2.2 RF Spectra-Static Response Relation in the General Case

While Eq. 6.11 is useful for interpreting and calculating RF spectra, as well as for a novel measurement, it is true only insofar as the RF spectral line shift is proportional to the density, up to an additive constant. Though this is exact in both the deep MI and SF limits, and to a good approximation true in between, it need not be accurate for general systems. Consequently, it is valuable to pursue an exact, general relation for those cases. I will still assume the sum rule-based calculation suffices.

The frequency is in general linearly related to $\mathcal{S} \overset{\text{def.}}{\equiv} \langle n(n-1)\rangle/\langle n\rangle$. So analogous to the previous calculation, I consider the spectra as a function of $\mathcal{S}$. I introduce the analog of Eq. 6.9:

$$C(\mathcal{S}) = -\int_{\mathcal{V}_\mathcal{S}} d\mathbf{r}\, n(\mathbf{r}).$$

The RF spectrum $\mathcal{I}(\mathcal{S})$ is

$$\mathcal{I}(\mathcal{S}) = -\frac{d}{d\mathcal{S}}\left[\int_{\mathcal{V}_S} d\mathbf{r}\, n(\mathbf{r})\right]$$

since $\mathcal{I}(\mathcal{S})\delta\mathcal{S} = C(\mathcal{S}) - C(\mathcal{S}+\delta\mathcal{S})$ from the definition of $\mathcal{I}$ and C. Differentiating, essentially repeating the steps to Eq. 6.10 while assuming the mean shift is monotonic in displacement from the trap center,

$$\mathcal{I}(\mathcal{S}) = -\gamma\frac{d\mu_{\text{eff}}}{d\mathcal{S}} r(\mathcal{S})n(r(\mathcal{S})). \tag{6.12}$$

This is the generalization of the RF spectrum-compressibility relation: $\frac{d\mathcal{S}}{d\mu_{\text{eff}}}$ takes the role of the compressibility, and $n(r(\mathcal{S}))$—that is, the density corresponding to $\mathcal{S}$—takes the role of the density ρ.

Finally, as in the approximate case previously treated, the response $\frac{d\mathcal{S}}{d\mu_{\text{eff}}}$ can be related to number fluctuations through a usual "correlation-response" derivation:

$$\frac{d\mathcal{S}}{d\mu_{\text{eff}}} = \beta\sum_j\left[\frac{\langle n_i(n_i-1)n_j\rangle}{\langle n_i\rangle} - \langle n_i(n_i-1)\rangle\right].$$

6.5.2.3 Summary

In summary, in Chap. 4, we previously thought of the RF spectral peaks as coming from corrugated density profiles, and the height of a RF spectrum getting larger as the density changes more slowly with μ or r. Here, we made this precise, and showed (in an approximation) that additionally (1) the RF spectral weight corresponding to density n is related in a precise way—Equation 6.11– to the compressibility and n; and (2) this is directly related to the number squeezing and fluctuations of the state. We then derive the general versions of this statement, Eq. 6.12. Finally, we pointed out that this opens up a new window into the local thermodynamic and quantum phases that appear in a inhomogeneous, trapped system.

The relations readily generalize to arbitrary dimension.

6.6 Determining Density Profiles from RF Spectra

Here we show that the RF spectrum diagram can be uniquely inverted to yield the density profile up to an overall spatial scale, under the assumptions of a monotonically changing density in a harmonic trap.

6.6.1 Determining Density Profiles from Spectra When Mean Energy Shift is Proportional to the Density Plus a Constant

First, we use the development of Sect. 6.5.2, and work in the same approximation as there: that the RF spectral shift is proportional to the density up to some constant-in-the-trap offset. Later we will examine how this approximation can be eliminated

Equation 6.10 shows that the RF spectrum is given by

$$\mathcal{I}(\rho) = -\gamma\rho r^2 \frac{dr}{d\rho}$$

where γ is a constant independent of x and ρ. This is a differential equation for ρ, written in the usual form as

$$\frac{d\rho}{dr} = -\frac{\gamma}{\mathcal{I}(\rho)} r^2 \rho.$$

Re-arranging, and always considering equality *up to a constant*,

$$\frac{1}{r^2\rho}\frac{d\rho}{dr} = -\frac{1}{\mathcal{I}(\rho)}$$
$$\Leftrightarrow \frac{d(\log\rho)}{d(r^3)} = -\frac{1}{\mathcal{I}(\rho)}.$$

In this form, it is clear that any constants can be absorbed into r, so that constants only change the overall spatial scale. Since this is one-dimensional differential equation, we can integrate to obtain

$$-r^3 + r_0^3 = \int_{\log\rho(r_0)}^{\log\rho} d(\log\rho)\mathcal{I}(\rho)$$
$$-r^3 = \int_{\log\rho(r_0)}^{\log\rho} d\zeta\,\mathcal{I}(e^\zeta) - r_0^3.$$

The right-hand side of this equation defines a new function

$$F(\rho) \overset{\text{def.}}{\equiv} \int_{\log\rho(r_0)}^{\log\rho} d\zeta\,\mathcal{I}(e^\zeta) \tag{6.13}$$

which can be calculated and inverted numerically to obtain ρ as a function of r. Thus

$$\rho = F^{-1}(-r^3) \tag{6.14}$$

and the full density profile can be obtained by the RF spectrum through Eqs. 6.13 and 6.14. In regions where the spectrum is essentially flat, the spatial variation is correspondingly e^{-r^3}. Note that we need to know $\rho(r_0)$ for some arbitrary r_0 as a reference. We can use the wings where the cloud is non-interacting as a reference if necessary where $\rho(r_0)$ is calculable.

Actually, as discussed in Sect. 6.5.2, the expressions above only holds for $\mathcal{I}(\rho+f)$ where f is an offset, and not for $\mathcal{I}(\rho)$ itself, so our density profiles may be shifted by a constant value by the values we will determine below.

6.6.2 *Determining Density Profiles from RF Spectra in the General Case*

The general case, where the RF spectrum energies is not simply proportional to the site filling plus a constant, was considered in Sect. 6.5.2, as well. We can try to invert this relation in a similar manner as in the previous subsection by noting that the spectrum is given by (equation is up to a rescaling of r again)

$$\mathcal{I}(\mathcal{S}) = -n(\mathcal{S})r^2\frac{dr}{d\mathcal{S}}$$

which is re-arranged to give

$$\frac{d\mathcal{S}}{dx} = -r^2\frac{n(\mathcal{S})}{\mathcal{I}(\mathcal{S})} \tag{6.15}$$

which is not solvable without knowledge of $n(\mathcal{S})$. This is the general formula for determining the density profile from the RF spectrum. In practice, if one knows *something* about the relation between $\mathcal{S}$ and n, one can hope to solve this equation and obtain the $\mathcal{S}$ profile in the trap. As an example, assume that one knows $\mathcal{S} = n + Dn^2$ for some constant D, so one knows that $n(\mathcal{S}) = (1/2D)\ (-1 \pm \sqrt{1+4D\mathcal{S}})$. Then one can solve Eq. 6.15 if one knows D; even if D is unknown, the differential equation can be solved for all possible values of D and all possible density profiles examined that could correspond to the given spectrum.

These equations can be readily generalized to (possibly anisotropic) harmonic traps in arbitrary dimension.

6.6.3 *Cooling and Quantum Register Initialization*

A key requirement for all quantum algorithms is the production of a well-characterized initial ("fiducial") state. A popular choice for neutral ultracold atomic gas proposals is the n = 1 MI. However, forming the MI invariably forms defects. Spectroscopically addressing and removing atoms from sites having too many, the system moves towards an ideal MI state, as suggested in Ref. [10]. We point out that such manipulation towards a single quantum state constitutes cooling. This should then enable temperatures even lower than present. The MI can then sympathetically cool others.

6.6.4 *Measuring Spatial Correlations; Quantum Computer Readout*

As a final application of RF spectroscopy techniques, we show how it can be used to measure spatial correlations in more details than other techniques. Actually, we present two techniques: first, an idea with limited scope, and then a more interesting idea that enables substantially informative imaging of spatial correlations. Specifically, it provides the ability to measure all m-point density correlation functions $\langle n_i n_j \cdots n_m \rangle$ or equivalent, the probability distribution $P(n_i, n_j, n_k, \ldots)$ for having n_i atoms at site i, n_j at site j, etc. We mention in passing that this might function as a state readout for a quantum computer, since it can provide the probability of two qubits i and j being in states s_i and s_j respectively. We will show elsewhere that this enables partial measurement of the n-body Green's functions of *arbitrary* systems for n not too large.

Very recent developments—after the writing of this thesis—allow one to directly measure densities with single site resolution in certain experiments. By doing many shots, one can in principle map out $P(n_i, n_j, n_k)$. Nevertheless, this requires an extreme amount of averaging and is available in a very limited set of apparatuses, so RF spectra may be beneficial in practice.

6.6.4.1 Background for Proposed Experimental Probe: Energy Shifts in the Presence of Long-Range Interactions

The method I propose is requires a small, off-site interaction potential, different for *a-a* and *a-b* interactions. The system is tuned deep into the MI state, and the long range interaction needs to be small enough so as not to disturb the state of the system. Henceforth, we refer to the off-site potential as "long-range." Possible candidates for the long-range interaction then are dipolar condensates and nearest-neighbor interactions. Condensates with significant dipole moments have been realized with chromium, [12] with stronger dipolar interactions via diatomic molecular condensates or using Rydberg atoms [13].

Because our proposed spatial correlation measurement will not depend in any crucial way on the details of the potential used, we consider a general 2-body potential

$$V_l \stackrel{\text{def.}}{\equiv} \sum_{i \neq j} \sum_{\alpha,\beta} V^{(l)}_{\alpha\beta}(i,j) n_{i,\alpha} n_{j,\beta}$$

where $V^{(l)}_{\alpha\beta}(i,j)$ give the interaction of an atom of type α at site i with an atom of type β at site j, the first sum runs over all pairs (i,j) such that $i \neq j$, and the second runs over $\alpha, \beta \in \{a, b\}$. The potential satisfies $V^{(l)}_{\alpha\beta} = V^{(l)}_{\beta\alpha}$.

For simplicity of analysis, we consider the probe system to be confined to one-dimension; this is not a necessity for the measurement technique, but simplifies the exposition of the idea behind the technique.

6.6.4.2 RF Spectra with Long-Range Interactions

In the deep MI, since the sites decouple, the complete set of eigenstates is $\left|\{n_i\}\right\rangle \stackrel{\text{def.}}{\equiv} \bigotimes_i \left|n_i\right\rangle_i$ where $\left|n\right\rangle_i$ is the state with n particles at site i. As discussed previously, a system in an incoherent mixture of initial eigenstates $|\psi\rangle$ with probability P_ψ responds with probability P_ψ if the laser is on resonance with the CS energy for state ψ. In other words, as discussed in the sections on finite temperature, in calculating the spectra the expectation values occurring in the formulas for the RF spectra are not taken over a thermal ensemble of quantum states – this would give the "center" of the entire CS diagram. Instead, the contribution from each eigenstate is calculated using the quantum expectation in that eigenstate and these are summed, weighted with the probability P_ψ of that eigenstate.

The RF spectral shift for the state $|\{n_i\}\rangle$ is, using the method of calculating commutators from Ref. [3],

$$\begin{aligned} \Delta_j = \delta E_{1,j} + (U_{ab} - U_a)\left(n_{j,a} - 1\right) \\ + 2 \sum_{i \neq j} \delta V_{i,j} n_{i,a} \end{aligned} \tag{6.16}$$

with $\delta V_{i,j} \stackrel{\text{def.}}{\equiv} V^{(l)}_{ab}(i,j) - V^{(l)}_{aa}(i,j)$.

Notice the crucial feature of the resulting spectral shifts: Equation 6.16 shows that each spatial arrangement of the n_i's leads to a different Δ_j: **each spectral peak is *uniquely* associated with a spatial correlation**, and allows us to determine the probability of that configuration occurring in the initial state. That is, the on-site contribution to the energy shift is the number of particles at site j times $U_{ab} - U_a$, while the number of particles at the neighboring site times δV_{ij} (which should be somewhat smaller than $U_{ab} - U_a$ to unambiguously associate peaks with spatial variations—see Sect. 6.6.4) for i and j being nearest neighbors gives the nearest neighbor contribution, and so on. The weight of the peak is proportional to the number of sites in the transfer laser's path satisfying the criterion that the

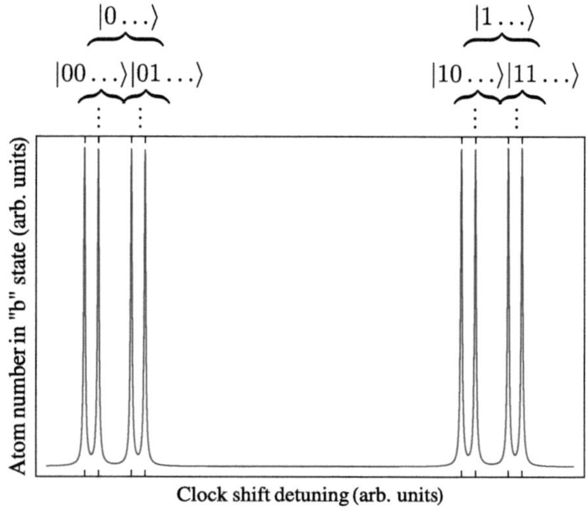

Fig. 6.4 Deep Mott RF spectra with dipolar interactions for the state given in the text. The correspondence between peak and spatial variation is indicated, where a string "$n_1n_2n_3$" indicates n_1 particles in the first neighbor site, n_2 in the second neighbor, and n_3 in the third neighbor. The labels instruct how to read-off the probabilities for each state

correlations in its nearest neighbors, next-nearest neighbors, and so on are appropriate to the energy of the transfer laser.

Figure 6.4 shows an example CS diagram and how to read off the spatial correlations. It is generated, just for illustration, with a three-(relevant)-site MI with all possible states such that each site is unoccupied or singly-occupied with equal probability. We have labeled some of the peaks to visually illustrate how the probabilities corresponding to various number fluctuations/correlations are manifested in the CS diagram.

As a final note, if one can address a single lattice site, one can measure the correlations specifically involving that site; otherwise one measures the average over the transfer laser beam's volume.

6.6.4.3 Constraints on Experimental Parameters

Let's thoroughly examine the restrictions on the interaction due to competing requirements of not disturbing the MI and having an energy shift large enough to be resolvable. Additionally, the interaction needs to be small enough that the peaks can easily be associated with spatial variations—for too large interactions, the peaks of Fig. 6.4 get mixed up and are difficult to assign to their given spatial number configurations. The condition that the MI not be disturbed by the long-range interaction is

$$V_{\alpha\beta}^{(l)}(i,j) \ll U_a \tag{6.17}$$

for all i, j; this is due to the fact that U_a sets the energy scale for the Mott physics. This condition is not terribly difficult to satisfy; if nothing else, one can increase the lattice depth and hence U_a.

A helpful condition for associating particular peaks with particular states is the condition that

$$\left[V_{ab}^{(l)}(i,j) - V_{aa}^{(l)}(i,j)\right](n-1) < U_{ab} - U_a \tag{6.18}$$

is the difference in energy shifts between states with n and $n+1$ atoms, for n the maximum filling of interest to which the system fluctuates (more attention to the i,j dependence is required and will be given in a moment). This ensures that the shifts due to interactions with neighboring sites is less than that due to the on-site CS, so that all the side-peaks to a given filling n (due to offsite correlations) are less than those of filling $n+1$ and one can easily read off the correlations.

In competition with the previous conditions, however, is the condition that the correlations between i and j be resolvable; this requires that

$$V_{ab}^{(l)}(i,j) - V_{aa}^{(l)}(i,j) > W \tag{6.19}$$

where W is the linewidth of the MI peaks or frequency resolution, whichever is larger. It is not too difficult to meet these requirements since W and $(U_{ab} - U_a)/n$ can be separated by orders of magnitude: they differ already factor of 2-3 in Campbell et al's experiments, while using, say, sodium would increase $U_{ab} - U_a$ by a factor of 30 [10]. The linewidth of Ref. [10] is probably pulse limited, in which case longer pulses could decrease W further. Conservatively, a factor of roughly 200 between $U_{ab} - U_a$ and W seems achievable. If one can find or design an interaction such that $V_{ab}^{(l)} - V_{aa}^{(l)}$ is close to the minimum allowed value of W, one can measure fluctuations in particle number up to 200—likely a far larger number than will be of interest in the near future.

Actually, so far we have failed to treat the interaction's i,j dependence carefully, and doing so imposes further constraints. Because the dipole-dipole interaction, which decays as $1/r^3$, is the most favorable to measuring correlations between far-away sites in neutral atomic systems, we will give it special consideration in the analysis. The constraint given by Eq. 6.17 is easily satisfied.

Turning to the constraint of Eq. 6.18 with a proper consideration of i, j dependence, it should more properly state that, picking a site j, the sum over all $i \neq j$ of the interaction with that site gives an energy that is smaller than the energy shift between fillings n and $n+1$ at site j; the sum was ignored in this formula. However, even for the longest range potential of interest, the dipole-dipole-interaction, this sum is of the form $V\sum_{l=1}^{\infty} n_l/l^3$ for some constant V with l the displacement away from j. We can get an upper bound on this energy by considering the maximum filling to which the system might fluctuate, say $n_{\max}$ and replace n_l by this filling, yielding for the sum $Vn_{\max}\zeta(3) \approx 1.2Vn_{\max}$ where ζ is the

Riemann zeta function. Thus, we need only replace $U_{ab} - U_a$ with $(U_{ab} - U_a)/1.2$ in Eq. 6.18.

Equation 6.18 needs further augmentation for sideband resolvability: all of the $|n_1, n_2, \ldots\rangle$ states must be lower in energy from $|n_1, n_2, \ldots\rangle$ states, and so on for states with the first m-sites with equal filling. This limits one to $1.2n_{\max} \lesssim n_{\max} + 1$, which is satisfied when $n_{\max} \lesssim 5$. Faster decaying potentials enable larger $n_{\max}$, although that comes with another tradeoff (see next paragraph).

Finally, we include i, j dependence in Eq. 6.19's constraint. This equation is easily corrected: it just says that for any pair of sites i and j's whose correlations are to be measured, one must have $V_{ab}^{(l)}(i,j) - V_{aa}^{(l)}(i,j) > W$. The correlations between sites i and j will be resolved for sites (and only for sites) such that Eq. 6.19 is satisfied. For the dipole-dipole potential, we see that this constraint simultaneous with the constraint in Eq. 6.18 serious limits the distance between sites whose correlations can be measured, if one wants also to resolve the nearest-neighbor correlations: we want to maximize $V_{ab}^{(l)}(i,j) - V_{aa}^{(l)}(i,j)$ while satisfying Eq. 6.18 Call the resulting optimal nearest neighbor interaction V. Then for sites $2, 3, 4, \ldots$ sites apart, the energies are $V/8, V/27, V/64, \ldots$. Hence, even if $V/W \sim 100$, only sites a few lattice spacings apart are measurable, and at that distance, rather than being able to measure particle fluctuations ~ 100 as in the nearest-neighbor correlations, one is limited to fluctuations that are on the order of unity. By increasing $U_{ab} - U_a$ and V correspondingly, one can do better, but the growth in distance between sites measurable is slow, going as $(U_{ab} - U_a)^{1/3}$.

We have just seen that the constraints in Eqs. 6.17, 6.18, and 6.19 strongly constrain measuring correlations between distant sites simultaneously with the nearest sites, but they provide ample opportunity to measure at least the nearest-neighbor correlations. Moreover, even if we use only the nearest neighbors, arbitrary distances can be created between the two sites simply by adjusting the lattice spacing of the optical lattice. Hence the only feature and constraint uncovered by proper treatment of the i,j dependence, then, is the inability to easily measure correlators involving more than two sites. Through careful choice of interactions to fully exploit the range of energies allowed by Eqs. 6.17, 6.18, and 6.19, by straightforward development of present apparatuses three- or four-site n-body correlations (for fairly large n) could be measured!

6.6.4.4 Measuring Away from the Deep Mott State

Let's consider the system *away from the deep Mott limit*: it may be at finite temperature in the MI, at finite t/U in the MI, or in the SF. More generally, other interactions/Hamiltonian terms could be turned on to study new phases (for example, one could engineer dipolar interactions, spin-dependent interactions, more complex lattices, ...); we show here that all these systems can be measured by straightforward application of the method suggested in Sect. 6.6.4.

In Sect. 6.6.4 we showed that RF spectra diagrams with perturbing state-dependent off-site interactions measure probabilities of spatial variations, assuming the system was in the deep MI. However, for a general boson lattice system, it is in principle easy to imprint the correlations of the system into an out-of-equilibrium deep MI, simply by ramping up the lattice instantaneously.

Recall from elementary quantum mechanics the so-called "instantaneous" or "sudden" approximation. The Hamiltonian of the system is switched from H_- to H_+ instantaneously at a time $t = 0$. Let's call the state for $t<0$, $|-,t\rangle$, and call $|S\rangle$ the eigenstates of H_+. The key observation is that the wavefunction is continuous in time, so that enforcing continuity and expanding the final state in terms of the eigenstates of H_+ gives

$$\lim_{t \to 0^-} |-,t\rangle = \sum_S A_S |S\rangle,$$

with $A_S = \langle S|-,0\rangle$. For $t > 0$, $|A_S|^2$ is the S-state probability.

This is usefully combined with Sect. 6.6.4's measurement. We work with a general system (most commonly, but not necessarily) in equilibrium for a Hamiltonian H_m of interest, which may be arbitrarily complicated; then the Hamiltonian is suddenly switched to that of the deep MI limit of the BH Hamiltonian. Via the sudden approximation, the system is in the equilibrium state of H_m, but evolves under the Hamiltonian for the deep MI limit of the BH model. The method of Sect. 6.6.4 can then be utilized to measure the correlations of this state.

An illuminating, simple, and very important example is provided by the SF in the BH model. Initially the system is in equilibrium in a SF region of the phase diagram. Then the lattice may be ramped up quickly; the site-to-site hopping is suppressed (exponentially) with the increasing lattice height, and hence one very rapidly enters the deep MI. As in the general case, the correlations are measurable via the procedure of Sect. 6.6.4.

In short, we have proposed an experimental protocol for measuring spatial correlations. The requisite technology is mostly developed: the components of the measurement technique—RF spectra, beyond on-site interactions, and dynamic control of experimental parameters – have all been demonstrated.

6.7 Conclusions

The finite temperature RF spectra in the superfluid and Mott insulator were considered. Both the superfluid and Mott insulator were shown to develop qualitatively new features at sufficiently high temperatures: in the superfluid, a peak near zero frequency comes from the dilute tail of excitations in the cloud, and in the Mott insulator a set of peaks occurs that correspond to fillings absent in the ground state.

Furthermore, we showed that finite temperature *homogeneous* Mott states or non-equilibrium states can lead to multiple Mott peaks, since different filling eigenstates are occupied: this gives an illusion of inhomogeneity if interpreted using the na zero-temperature trap theory. In a trap, we saw that a $n=1$ zero temperature Mott insulator can appear like a series of Mott shells in the RF spectrum. This demonstrates another reason that this technique is not sufficient to conclusively identify the Mott state.

Finally, we propose and analyze a number of experimental applications: thermometry, cooling, measuring density profiles from the RF spectrum, measuring the number-squeezing correlations of the superfluid, and measuring spatial correlations of the system, each of which significantly improves important experimental capabilities.

To overview the applications, the proposed *thermometry* allows measurements temperatures down to the smallest Mott gaps one can create, which are in turn limited only by how small one can make the hopping; since wider lattice depths decrease this exponentially, this appears to offer considerable opportunity. We have argued that the thermometry possesses many additional desirable features of accuracy, robustness, and self-diagnosis lacking in current methods. The *cooling* offers prospects to reach lower temperatures. The *inversion of RF spectra* to give density profiles gives complementary information to Abel-transforming columnar integrated density profiles. A similar argument shows that in some approximation the magnitude of the RF spectrum is related in a simple fashion to the *compressibility*. Finally, the most ambitious proposal is *an experimental technique to measure general n-point density correlation functions*.

References

1. T.-L. Ho, Q. Zhou, Phys. Rev. Lett. **99**, 120404 (2007)
2. D. Jaksch, C. Bruder, J.I. Cirac, C.W. Gardiner, P. Zoller, Phys. Rev. Lett. **81**, 3108 (1998)
3. M. Öktel, T.C. Killian, D. Kleppner, L.S. Levitov, Phys. Rev. A **65**, 033617 (2002)
4. C.J. Pethick, H.T.C. Stoof, Phys. Rev. A **64**, 013618 (2001)
5. K.R.A. Hazzard, E.J. Mueller, Phys. Rev. A **81**, 033404 (2010)
6. K.R.A. Hazzard, E.J. Mueller, Phys. Rev. A **76**, 063612 (2007)
7. M.Ö. Oktel, L.S. Levitov, Phys. Rev. Lett. **83**, 69 (1999)
8. C.J. Pethick, H. Smith, *Bose-Einstein condensation in dilute gases*. (Cambridge University Press, The Edinburgh Building, Cambridge CB2 2RU, UK 2001)
9. D.A.W. Hutchinson, E. Zaremba, A. Griffin, Phys. Rev. Lett. **78**, 1842 (1997)
10. G.K. Campbell, J. Mun, M. Boyd, P. Medley, A.E. Leanhardt, L.G. Marcassa, D.E. Pritchard, W. Ketterle, Science **313**, 649 (2006)
11. G.G. Batrouni, V. Rousseau, R.T. Scalettar, M. Rigol, A. Muramatsu, P.J.H. Denteneer, M. Troyer, Phys. Rev. Lett. **89**, 117203 (2002)
12. A. Griesmaier, J. Werner, S. Hensler, J. Stuhler, T. Pfau, Phys. Rev. Lett. **94**, 160401 (2005)
13. T. Lahaye, C. Menotti, L. Santos, M. Lewenstein, T. Pfau, Rep. Prog. Phys. **72**, 126401 (2009)

Chapter 7
RF Spectra: Summary, Conclusions, and the Future

Summarizing our studies of RF spectroscopy, we have showed that two common but qualitatively distinct pictures of RF spectroscopy—one based on sum rules, and one based on single particle Green's functions—emerge as special limits of a more complete calculation, depending on the final state interactions through the quantity $(U_{ab} - U_{aa})/t_a$. In the deep Mott and superfluid limits, we have extended our calculations to finite temperature. We have also discussed how RF spectra with interactions beyond on-site may be used to probe full joint number distributions on multiple sites; candidates with appropriate interactions are Rydberg excitations or molecular gases.

Experimentally, an exciting future goal for the field would be the observation of the bimodal spectrum identified in Chap. 5. The main impediment to immediate observation by the MIT group is that they use ^{87}Rb, which in typical optical lattices displays a significant bimodal feature in only a small part of the phase diagram. Two routes may circumvent this: (i) one may use spin dependent optical lattices to displace the two spin components' lattices, in which case U_{ab} can be significantly smaller than U_{aa} [this setup is being pursued by the MIT group (W. Ketterle, 2007, private communication)] or (ii) one may use a different species, such as Na, Yb, or Cs, which possess larger $U_{ab} - U_{aa}$'s or enable Feshbach resonances to be used (Cheng Chin's group is ideally poised to study this using Cs and high resolution *in situ* imaging techniques for the 2D Bose–Hubbard model). Further exploration could lead naturally to observation of non-quasiparticle behavior in the quantum critical regime using the tools developed in Chap. 13.

Theoretically, it is desirable to better understand the structure beyond the flat-band approximation inherent in the Gutzwiller approximation (that is, to incorporate intermediate range correlations). There exists a formalism to go beyond Gutzwiller in a way that correctly reproduces the Boguliobov low energy excitations in the deep superfluid [1, 2]. Konabe et al. and Ohashi have computed the RF spectra for vanishing final state interactions ($U_{ab} = 0$) using this method [3, 4]. This approach captures the finite bandwidth in the Mott insulator as well as

K. R. A. Hazzard, *Quantum Phase Transitions in Cold Atoms and Low Temperature Solids*, Springer Theses, DOI: 10.1007/978-1-4419-8179-0_7,

the different dispersion for the a and b type atoms in the deep superfluid limit. Finally, the latter limit could be treated simply and directly via Boguliobov theory.

As another theoretical avenue, the prospect of a more complete understanding of the finite temperature phase diagram is highly exciting. As one example, discussed in Chap. 13, the RF spectra can display non-quasiparticle behavior. To capture the quantum dynamics in the critical region, we will require a different approach. The large-N and ϵ expansions offer possibilities to qualitatively capture the physics [5]. Cold atom experiments have potential to dramatically improve our understanding of the validity of these expansions.

Finally, an understanding of how the state created in RF spectroscopy evolves at long times to equilibrium would be interesting from the point of view of far from equilibrium physics, for understanding RF spectra in practice (which are only moderately in the linear response regime), and could shed light on the accuracy of atomic clocks [6, 7]. Numerically solving our time-dependent mean field theory beyond the linear response regime would be a natural building block for a theory of this equilibration, but the extent to which this provides an accurate treatment of the physics is unclear.

Another interesting idea to calculate quantum critical properties (more generally than RF spectra) is inspired by our work: we can try to develop a perturbation theory in $\eta = (U_{ab} - U_{aa})/t_a$. This may be feasible since for $\eta = 0$, one obtains a simple spectrum of an unshifted delta function, and for small η in our calculation this always leads simply to a shift. It would be interesting to see if this approach allows a seemingly quite novel alternative to other types of expansions, with the additional advantage that the $\eta = 0$ limit is physical, in contrast to large-N and ϵ expansions.

References

1. C. Menotti, N. Trivedi, Phys. Rev. B **77**, 235120 (2008)
2. K. Sengupta, N. Dupuis, Phys. Rev. A **71**, 033629 (2005)
3. S. Konabe, T. Nikuni, M. Nakamura, Phys. Rev. A **73**, 033621 (2006)
4. Y. Ohashi, M. Kitaura, H. Matsumoto, Phys. Rev. A **73**, 033617 (2006)
5. S. Sachdev, Quantum Phase Transitions. (Cambridge University Press, Cambridge, 2001)
6. S. Gibble, S. Chu, Phys. Rev. Lett. **70**, 1771 (1993)
7. C. Fertig, K. Gibble, Phys. Rev. Lett. **85**, 1622 (2000)

Chapter 8
Rotation, Inducing Gauge Fields, and Exotic States of Matter in Cold Atoms

Rotating quantum systems leads to dramatic and exotic phenomena. The first rotational phenomena to be experimentally explored in cold atoms was vortices and vortex lattices in a rotating superfluid. Because a superfluid is irrotational, all rotation is contained in topological defects with circulation quantized in units of $\hbar$. This is a phenomena that occurs in superfluid ^{4}He and ^{3}He, as well as in type-2 superconductors (superconductors only show vortices in a certain limit, corresponding to type-2 superconductors, because the electromagnetic field energy has to be considered in addition to the electron fluid [1–3]). Images of vortices vortices in cold gas experiments directly showed this spectacular manifestations of superfluidity.

Faster rotation is predicted to manifest more exotic behavior. A specific case is discussed in Sect. 8.1, but we can understand this very generally from a few qualitative perspectives. One way of looking at the physics is that a gauge field, as is introduced by rotation, introduces a quantum mechanical phase for a particle to traverse a loop. This introduces or increasing the severity of the "sign problem" into the physics, making it difficult to simulate. Another perspective is that the wavefunction must be single-valued, yet minimizing the energy cost of phase twists upon traversing a loop fails to satisfy this condition; this "frustration" leads any loop drawn in the system to have nearly degenerate configurations, a competition which can be resolved by interactions. Perhaps the most general viewpoint is the observation that rotation causes the appearance of a degeneracy. This is particularly striking for fast rotation when rotation leads to macroscopically degenerate Landau levels [4].

More generally than just rotation, we are interested in inducing "effective gauge fields." I will discuss methods of stabilizing exotic states in cold atom systems, especially those induced by "effective" gauge fields. Most often, one engineers the system so that the neutral particles behave as if they were atoms in a homogeneous magnetic field. Rotation is the most common approach [5, 6], but there has been recent success with light-induced gauge potentials—Refs. [7–9] give some

K. R. A. Hazzard, *Quantum Phase Transitions in Cold Atoms and Low Temperature Solids*, Springer Theses, DOI: 10.1007/978-1-4419-8179-0_8,

prominent proposals and Ref. [10] gives an experimental realization. These techniques also enable non-Abelian gauge fields [11] and may allow creation of the stronger gauge fields necessary to induce exotic behavior (more on this below). The most interesting possibility is the study of topological phases of matter (e.g., the fractional quantum hall effect). I will discuss some of these approaches.

It is worth noting that the gauge fields discussed here are static, external gauge fields. Emulating dynamic gauge fields with their own degrees of freedom is another, further off, experimental possibility for which theoretical proposals exist (for one example, see Ref. [12]).

This chapter provides the background to understand methods to induce gauge fields, as well as the properties of the resulting physical states.

8.1 Physics of Rotating Particles/Particles in Gauge Fields

Much fascinating physics occurs when particles are subjected to gauge fields. I briefly review some of the most intriguing and experimentally relevant examples: vortices, the integer quantum Hall effect and interplay with lattices (Hofstadter butterfly), and the fractional quantum Hall effect. Table 8.1 gives an overview of classes of physics formed by the interplay of rotation, lattices, and the trap for bosonic atoms in limits where one or two of these dominate the physics, demonstrating the richness of the general case, even for purely scalar bosonic systems.

Table 8.1 Behavior of rotating lattice bosons in the simplest limits for *homogeneous* systems. The various length scales characterize interactions (scattering length a), the lattice depth (lattice site's oscillator length ℓ), the lattice spacing (d), and rotation (rotational length ℓ_R)

Limit	Behavior
$(a=0)\wedge(\ell_R \gg \ell)\wedge(\ell \ll d)$	Hofstadter butterfly [13]
$(a=0)\wedge(\ell \gg d)$	Thouless butterfly (dual of Hofstadter butterfly) [14]
$(a \ll r_s)\wedge(\ell_R \gg \{d, r_s\})$	Vortex lattices [15–19]
$(a \gg \ell_R - \ell)\wedge(\ell_R \lesssim r_s)\wedge[(r_s \gg d)\vee(\ell \gg d)]$	Bulk fractional quantum hall [20–22]
$(a \gg \ell_R - \ell)\wedge(\ell_R \lesssim r_s \lesssim \ell \ll d)$	FQH puddle arrays [23–30]
$(\ell_R = \infty)\wedge(\ell \ll d)$	Non-rotating strongly interacting bosons: atomic SF/molecular SF QPT [31–33]
$(\ell_R = \infty)\wedge(0 < a \ll \ell \ll d)$	Non-rotating ordinary Bose-Hubbard model [34, 35]

Here $\wedge$ represents logical "AND" and $\vee$ the logical "OR." One sees that even for the homogeneous system there are a remarkable variety of limits: the ordinary Bose-Hubbard model is arguably the simplest of all the limits, and it already displays the highly non-trivial physics of the superfluid physics. Each of the limits is an areas of study in its own right. The references are representative, not comprehensive.

8.1.1 Vortices

Assuming interactions and rotation are weak enough so as to not destroy the Bose-Einstein condensate, we may describe it via a condensate wavefunction $\psi(\mathbf{r},t) = A(\mathbf{r},t)e^{i\phi(\mathbf{r},t)}$, writing ψ in terms of a real amplitude $A(\mathbf{r},t)$ and a real phase $\phi(\mathbf{r},t)$. Define the current operator $\mathbf{j}(\mathbf{r},t)$ via $\partial_t\rho = -\nabla\cdot\mathbf{j}(\mathbf{r},t)$ with $\rho(\mathbf{r},t) = |\psi(\mathbf{r},t)|^2$ the density at point $\mathbf{r}$ and time t. Then using that ψ solves the Schrödinger equation, we find that

$$\mathbf{j} = \frac{\hbar}{m}\nabla\phi. \tag{8.1}$$

Thus the rotational circulation is

$$\begin{aligned}\oint d\mathbf{r}\cdot\mathbf{j}(\mathbf{r}) &= \frac{\hbar}{m}\oint d\mathbf{r}\cdot\nabla\phi \\ &= \frac{\hbar}{m}(\phi_f - \phi_i)\end{aligned} \tag{8.2}$$

with ϕ_i the phase at $\mathbf{r}$ before traversing the loop and ϕ_f the phase at $\mathbf{r}$ after. If ϕ were a single valued function, these would be identical, but we require only that the wavefunction ψ is single valued, requiring ϕ to be quantized to $\phi = 2\pi n$ for integer n, and consequently

$$m\oint d\mathbf{r}\cdot\mathbf{j}(\mathbf{r}) = 2\pi\hbar n. \tag{8.3}$$

Thus the momentum circulation of a condensate is quantized in multiples of $h = 2\pi\hbar$. That is, it appears in discrete jumps, by vortex formation. Associated with the circulation, by contracting the loop, there must be a singularity in the wavefunction, at which point ρ vanishes, and this shows up as zero density points in the spatial absorption images, leading to spectacular images of vortices and vortex lattices in cold atoms [18].

8.1.2 Vortices in Lattices: Hofstadter Butterfly

Vortices interplay richly with external potentials, including static disorder, trapping potentials, and lattices. The first is responsible for pinning of flux lines in type-2 superconductors [3]; the second, in the case of a very thin toroidal potential (or a annular cell for, say, helium experiments), vortex tunneling across the annulus is expected to be the dominant decay mechanism for destruction of the superfluid flow [36, 37].

Externally imposed lattices potentials also display rich rotational physics. In a simple picture, one might expect pinning of low density cores at the maxima of potential energy in between sites. Some understanding exists for low vortex densities (compared to the lattice spacing). The picture is essentially true for fairly deep lattices and low vortex density, but for weak lattices the pinning may be destroyed because the vortex-vortex interaction favors a triangular lattice. This leads to a set of structural phase transitions, and the competition between phases may be selected by interactions. Goldbaum and Mueller have studied this, along with the effects of the trap [38, 39].

For high vortex densities, approaching one vortex per site, less is known. The non-interacting case is understood for very deep (tight-binding) and for very shallow (perturbative) lattices, and the energy spectrum shows a remarkable fractal structure as a function of vortex filling, called the "Hofstadter butterfly [13]."

There is similar structure for weak lattices [14]. As an aside, at some point I collaborated with an excellent undergraduate (Tim Yang) to numerically solve the equations for intermediate lattice depths, enabling us to study the crossover between these limits.

8.1.3 Quantum Hall Effect

Rotation rates faster than those considered above, so that the effect of rotation destroys the condensate, stabilizes exotic phases. In particular, various quantum Hall states can appear. It is impossible to characterize these states with a local order parameter. They possess topological order, undergo phase transitions without symmetry breaking, and display fractionalized excitations (See Chap. 9 and references therein). These highly intriguing states will be the focus of the rotational physics considered in this thesis, discussed in Chaps. 9 and 10 .

The simplest case occurs for non-interacting particles in the absence of a lattice, a continuum limit of the Hofstadter problem. The physics can be understood from a simple semiclassical picture of particles in a magnetic field (which Sect. 8.2.2 shows is equivalent to a rotating system). The quantum mechanical system in a strong magnetic field forms spin-polarized Landau levels, and when the chemical potential is between Landau levels in the bulk, the bulk is insulating. At the edge of the sample, the external confining potential increases. This decreases the effective chemical potential to cross the Landau level, so at the edge one obtains low energy excitations—from the velocity formula $\mathbf{v}_k = \partial\epsilon_k/\partial\mathbf{k}$, these are seen to be "chiral edge states" that circulate preferentially either clockwise or counter-clockwise around the edge of the system.

Even more interesting physics manifests for fractionally filled Landau levels. With a filling fraction ν we have a massive degeneracy corresponding to the many ways of distributing N particles among N/ν single particle states (Landau level states), and the ground state is then selected by the interactions. The contact interaction is most relevant for cold atoms, and in this case, in contrast to Coulombic interactions, the problem is exactly soluble. The exact solutions are the

Laughlin states, which were the variational states introduced by Laughlin for the Coulombic system [40].

Solution for a Contact Interaction. To construct the exact solution for a contact interaction, we first consider the solution for the non-interacting 2D homogeneous system's lowest energy eigenstates in a strong magnetic field (the first Landau level). These are all degenerate, with the symmetric gauge wavefunctions indexed by angular momentum k given by

$$\psi_k(x,y) = (\pi k!)^{-1/2} z^k e^{-z^* z/2}, \tag{8.4}$$

where $z = (x+iy)/d$ and $d = \sqrt{\hbar/(M\Omega)}$ with M the atomic mass [41].

The N-body wavefunction $\psi(z_1, z_2, \ldots, z_N)$ is a function of N complex numbers $z_j = x_j + iy_j$ describing the position of particle j. Since it can be written as a sum of products of the single-particle wavefunctions in (8.4), it is an analytic function of $z_1, z_2, \ldots, z_N$ times $e^{-\sum_j z_j^* z_j/2}$, ultimately a consequence of restricting the single particle Hilbert space to the lowest Landau level. It must satisfy the exchange symmetry $\psi \to \pm\psi$ upon exchanging any two of its arguments, with "+" for bosons and "−" for fermions. Since the Landau levels are degenerate, the interaction energy is the only energy in the problem, and thus the ground state minimizes the interaction energy. The minimum interaction energy possible is zero, and this is achieved when there is no overlap between pairs of particles. Thus wavefunctions of the form $\psi(z_1, z_2, \ldots, z_N) = \left[\prod_{i<j}(z_i - z_j)^{m_{ij}}\right] e^{-\sum_j |z_j|^2/2}$ with $m_{ij} > 0$ are all zero energy ground states. If a small energy penalty is added for angular momentum, slightly breaking the degeneracy of the Landau levels as is appropriate for a finite or a trapped system, one seeks to minimize the angular momentum, which is accomplished by $m_{ij} = 2$ for bosons and $m_{ij} = 1$ for fermions. This gives $\nu = 1/2$ and $\nu = 1/3$ Laughlin wavefunctions Ψ as the ground states, with

$$\Psi(\{z_j\}) = \left[\prod_{i<j}(z_i - z_j)^{1/\nu}\right] e^{-\sum_j |z_j|^2/2} \tag{8.5}$$

up to normalization. There is a nice physical interpretation: particles bind to a vortices. Excitations of this system may be constructed similarly.

There are several very interesting physical consequences of the ground and excited states: (1) the bulk has an energy gap, (2) there are gapless edge states, and (3) the gapped bulk excitations have fractional charge and statistics. One can also see that if Laughlin states of different fillings are stabilized (as occurs for non-contact interactions in semiconductor heterostructures) that transitions between them are indescribable via symmetry breaking and the phases must be characterized by some "topological order [37]".

8.2 Rotation

8.2.1 *Rotation and Angular Momentum Boosts—Principles and Experimental Techniques*

Consider a system described by a Hamiltonian H in some reference frame A. A system in a frame B rotating with frequency $\boldsymbol{\Omega}$—the vector accounts for the rotation axis—relative to reference frame A is described by the Hamiltonian

$$H' = H - \boldsymbol{\Omega} \cdot \mathbf{L} \tag{8.6}$$

where $\mathbf{L}$ is the total angular momentum operator of the system [41]. A consequence is that adding a fixed amount of angular momentum $\mathbf{L}$ to an angular momentum conserving system is equivalent to this, since the same description applies with $\boldsymbol{\Omega}$ a Lagrange multiplier to fix the total angular momentum.

Experimentally, one may rotate a whole system: one can rotate the entire trap, and lattice if one is there. Stably rotating lenses or lasers in unison is difficult, so a common technique is to shine light through a metal mask with holes cut in it, with the resulting diffraction pattern forming the lattice. Other techniques are also used (e.g., Ref. [42]). From the equivalence discussed above, another possibility is to impart angular momentum to the system by "stirring it." In such techniques, a focused laser beam generally provides the required spatially localized stirring potential.

8.2.2 *Rotation Leads to an Effective Magnetic Field (Gauge Field) and Limitations of Rotation for Inducing Exotic States*

I will derive how rotation leads to an effective magnetic field, considering the case of a single free particle. The derivation extends straightforwardly to multiple particles in an external potential.

A free particle is described by Hamiltonian $H = p^2/(2m)$. Equation 8.6 and the definition of angular momentum gives the rotating frame Hamiltonian to be

$$H' = \frac{p^2}{2m} - \boldsymbol{\Omega} \cdot (\mathbf{r} \times \mathbf{p}). \tag{8.7}$$

Cyclic invariance of the triple product yields

$$H' = \frac{p^2}{2m} - \mathbf{p} \cdot (\boldsymbol{\Omega} \times \mathbf{r}). \tag{8.8}$$

In principle one needs to worry about the non-commutation of the operators in the triple product, but here this is easily seen to be valid. We complete the square to give

$$H' = \frac{(p - m\boldsymbol{\Omega} \times \mathbf{r})^2}{2m} - m\Omega^2 r^2/2. \tag{8.9}$$

Thus, there is an added centrifugal force, and an effective gauge potential

$$\mathbf{A} = \frac{c\boldsymbol{\Omega} \times \mathbf{r}}{e} \tag{8.10}$$

which is precisely the gauge potential of a homogeneous magnetic field $B = mc\boldsymbol{\Omega}/e$ in the symmetric gauge [41]. Thus, neutral particles in a rotating frame will act exactly as charged particles in magnetic field (plus a harmonic centrifugal potential which may be absorbed into the harmonic trapping potential).

While rotation is can potentially lead to exotic states such as fractional quantum Hall states, it has severe limitations that have led cold atom physicists to propose new methods. In particular, to achieve fractional quantum Hall ground states the system must be rotated near the centrifugal limit where Ω matches the trap frequency ω. To be precise, by considering the competing ground states at nearby rotation frequencies, one finds that the maximum allowable frequency deviation to achieve fractional quantum Hall states scales as $|\Omega - \omega| \sim 1/N$ with N the number of particles. Thus, one requires high rotation frequency control and stability, severely limiting the size of fractional quantum Hall states that can be stabilized.

There are two main methods to circumvent this difficulty. (1) One method is to work with few-particle states so that $1/N$ is sufficiently large. These clusters remarkably show much of the physics of their thermodynamic brethren, even for $N \lesssim 10$, including gapped excitations that are well described by fractional charge and statistics, gapless edge modes, and a roughly quantized density [6, 16, 20–29, 43–45]. This is the approach we take in Chap. 9.

8.3 Other Methods of Inducing Gauge Fields

A number of alternative methods to create gauge fields have been proposed. These frequently rely on an atom receiving a momentum kick transverse to its motion from a laser when the atom hops [7, 8]. These ideas may also be extended to non-Abelian gauge potentials [11]. Erich Mueller and I have proposed and studied a particular class of these schemes, not discussed in this thesis.

Recently, these ideas have been generalized to spin-dependent gauge potentials, giving an effective spin orbit coupling (Ref. [46] gives one example). This has generated considerable excitement of the prospect of creating topological insulators in cold atoms. These are intriguing generalizations of the quantum Hall effect to systems without time-reversal symmetry breaking that have recently been observed in solid state systems (Ref. [47] gives a fairly up-to-date review). Perhaps most intriguingly, these system allow topological behavior to persist to three-dimensional systems, and generate two-dimensional conducting surfaces that are robust to weak disorder (in contrast to non-topological two-dimensional metals).

8.4 On-Site Correlations

Chapter 9 proposal to create fractional quantum Hall states leads to the creation of many few particle "puddles" of fractional quantum Hall states at each node of a two-dimensional optical lattice. This is a particular case of the more general problem of what happens in lattice systems when atoms within a site are allowed to correlate. Describing this requires going beyond the simple Bose-Hubbard description of Chaps. 2 and 10 describes a novel, efficient way of describing arbitrarily strong on-site correlations and estimates the magnitude of effects. Chapter 11 develops the theoretical tools to quantitatively calculate the model's parameters using path integral Monte Carlo simulations of moderate sized systems (the model can then be used to describe large, many-site systems).

References

1. J.R. Schrieffer, *Theory of Superconductivity*, Revised edn. (Perseus Books, New York, 1999)
2. P.G.D. Gennes, Superconductivity of Metals and Alloys. (Westview Press, USA, 1999)
3. M. Tinkham, *Introduction to Superconductivity*, 2 edn. (Dover Publications, New York, 2004) (Dover Books on Physics)
4. I. Bloch, J. Dalibard, W. Zwerger, Rev. Mod. Phys. **80**, 885 (2008)
5. N.R. Cooper, Adv. Phys. **57**, 539 (2008)
6. S. Viefers, J. Phys. Condens. Matter **20**, 123202, pp.14 (2008)
7. E.J. Mueller, Phys. Rev. A **70**, 041603 (2004)
8. A.S. Sørensen, E. Demler, M.D. Lukin, Phys. Rev. Lett. **94**, 086803 (2005)
9. M. Hafezi, A.S. Sørensen, E. Demler, M.D. Lukin, Phys. Rev. A **76**, 023613 (2007)
10. Y. Lin, R.L. Compton, K. Jimenez-Garcia, J.V. Porto, I.B. Spielman, Nat. Biotechnol. **462**, 628 (2009)
11. K. Osterloh, M. Baig, L. Santos, P. Zoller, M. Lewenstein, Phys. Rev. Lett. **95**, 010403 (2005)
12. H.P. Büchler, M. Hermele, S.D. Huber, M.P.A. Fisher, P. Zoller, Phys. Rev. Lett. **95**, 040402 (2005)
13. D.R. Hofstadter, Phys. Rev. B **14**, 2239–2249 (1976)
14. D.J. Thouless, M. Kohmoto, M.P. Nightingale, M. den Nijs, Phys. Rev. Lett. **49**, 405–408 (1982)
15. K.W. Madison, F. Chevy, W. Wohlleben, J. Dalibard, Phys. Rev. Lett. **84**, 806–809 (2000)
16. N.R. Cooper, N.K. Wilkin, J.M.F. Gunn, Phys. Rev. Lett. **87**, 120405 (2001)
17. P. Engels, I. Coddington, P.C. Haljan, E.A. Cornell, Phys. Rev. Lett. **89**, 100403 (2002)
18. J.R. Abo-Shaeer, C. Raman, J.M. Vogels, W. Ketterle, Sci. Agric. **292**, 476–479 (2001)
19. V. Schweikhard, I. Coddington, P. Engels, V.P. Mogendorff, E.A. Cornell, Phys. Rev. Lett. **92**, 040404 (2004)

20. N.K. Wilkin, J.M.F. Gunn, Phys. Rev. Lett. **84**, 6–9 (2000)
21. S. Viefers, T.H. Hansson, S.M. Reimann, Phys. Rev. A **62**, 053604 (2000)
22. J. Sinova, C.B. Hanna, A.H. MacDonald, Phys. Rev. Lett. **89**, 030403 (2002)
23. J.W. Reijnders, van F.J.M. Lankvelt, K. Schoutens, N. Read, Phys. Rev. Lett. **89**, 120401 (2002)
24. B. Paredes, P. Zoller, J.I. Cirac, Phys. Rev. A **66**, 033609 (2002)
25. N. Regnault, T. Jolicoeur, Phys. Rev. Lett. **91**, 030402 (2003)
26. U.R. Fischer, P.O. Fedichev, A. Recati, J. Phys B: At., Mol. Opt. Phys. **37**, S301 (2004)
27. T. Ho, E.J. Mueller, Phys. Rev. Lett. **89**, 050401 (2002)
28. N. Read, N.R. Cooper, Phys. Rev. A **68**, 035601 (2003)
29. M. Popp, B. Paredes, J.I. Cirac, Phys. Rev. A **70**, 053612 (2004)
30. S.K. Baur, K.R.A. Hazzard, E.J. Mueller, Phys. Rev. A **78**, 061608 (2008)
31. Y. Lee, Y. Lee, Phys. Rev. B **70**, 224506 (2004)
32. L. Radzihovsky, J. Park, P.B. Weichman, Phys. Rev. Lett. **92**, 160402 (2004)
33. M.W.J. Romans, R.A. Duine, S. Sachdev, H.T.C. Stoof, Phys. Rev. Lett. **93**, 020405 (2004)
34. M. Greiner, O. Mandel, T. Esslinger, T.W. Hänsch, I. Bloch, Nat. Biotechnol. **415**, 39 (2002)
35. D. Jaksch, C. Bruder, J.I. Cirac, C.W. Gardiner, P. Zoller, Phys. Rev. Lett. **81**, 3108 (1998)
36. I.M. Khalatnikov, An Introduction to the Theory of Superfluidity. (Westview Press, USA, 2000)
37. X. Wen, Quantum Field Theory of Many-body Systems: From the Origin of Sound to an Origin of Light and Electrons, reissue ed. (Oxford University Press, New York, USA, 2007)
38. D.S. Goldbaum, E.J. Mueller, Phys. Rev. A **79**, P021602 (2009)
39. D.S. Goldbaum, E.J. Mueller, Phys. Rev. A **79**, 063625 (2009)
40. R.B. Laughlin, Phys. Rev. Lett. **50**, 1395 (1983)
41. L.D. Landau, L.M. Lifshitz, *Quantum Mechanics Non-Relativistic Theory*, 3rd edn, vol. 3. (Butterworth-Heinemann, London, 1981)
42. R.A. Williams, J.D. Pillet, S. Al-Assam, B. Fletcher, M. Shotter, C.J. Foot, Opt. Express **16**, 16977 (2008)
43. N.K. Wilkin, J.M.F. Gunn, R.A. Smith, Phys. Rev. Lett. **80**, 2265 (1998)
44. N.R. Cooper, N.K. Wilkin, Phys. Rev. B **60**, R16279 (1999)
45. B. Paredes, P. Fedichev, J.I. Cirac, P. Zoller, Phys. Rev. Lett. **87**, 010402 (2001)
46. N. Goldman, I. Satija, P. Nikolic, A. Bermudez, M.A. Martin-Delgado, M. Lewenstein, I.B. Spielman, Phys. Rev. Lett. **105**, 255302 (2010)
47. M.Z. Hasan, C.L. Kane, Rev. Mod. Phys. **82**, 3045 (2010)

Chapter 9
Stirring up Fractional Quantum Hall Puddles

9.1 Chapter Overview

We theoretically explore the generation of few-body analogs of fractional quantum Hall states. We consider an array of identical few-atom clusters $n = 2, 3, 4$, each cluster trapped at the node of an optical lattice. By temporally varying the amplitude and phase of the trapping lasers, one can introduce a rotating deformation at each site. We analyze protocols for coherently transferring ground state clusters into highly correlated states, producing theoretical fidelities (probability of reaching the target state) in excess of 99%.

9.2 Introduction

Cold atom experiments promise to produce unique states of matter, allowing controllable exploration of exotic physics. For example, since rotation couples to neutral atoms in the same way that a uniform magnetic field couples to charged particles, many groups are excited about the possibility of producing analogs of fractional quantum hall states [1–15]. In particular, if a two dimensional harmonically trapped gas of bosons is rotated at a frequency Ω sufficiently close to the trapping frequency ω, then the ground state will have vortices bound to the atoms—an analog of the binding of flux tubes to electrons in the fractional quantum hall effect. The ground state will be topologically ordered and possess fractional excitations. Technically, the difficulty with realizing this goal experimentally has been

This chapter was adapted from "Stirring trapped atoms into fractional quantum Hall puddles" by Stefan K. Baur, Kaden R. A. Hazzard, and Erich J. Mueller, published in Physical Review A **78**, 061608(R) (2008). Stefan K. Baur and Kaden R. A. Hazzard contributed equally to this work.

K. R. A. Hazzard, *Quantum Phase Transitions in Cold Atoms and Low Temperature Solids*, Springer Theses, DOI: 10.1007/978-1-4419-8179-0_9,

that it requires Ω to be tuned to a precision which scales as $1/n$, where n is the number of particles. Responding to this impediment, several authors [9–15] have proposed studying clusters with $n \lesssim 10$. Such clusters possess many of the features of a bulk quantum hall liquid, and producing them would be a great achievement. Here we propose and study protocols for producing strongly correlated clusters of rotating atoms.

The issue prompting this investigation is that in such clusters there are very few mechanisms for dissipating energy, and hence experimentally producing the ground state of a rotating cluster is nontrivial. First, the small number of particles results in a discrete spectrum, and leaves few kinetic paths. Second, in the strongly correlated states of interest the atoms largely avoid each other, further blocking the kinetics. On these grounds, one should not expect to be able to cool into the ground state. Instead we advocate a dynamical process where one coherently *drives* the system into the strongly correlated state through a well-planned sequence of rotating trap deformations. This approach is based upon an analogy between the states of these atomic clusters, and the energy levels of a molecule. By deforming the harmonic trap, and rotating the deformation, one couples the many-body states in much the same way that an oscillating electric field from a laser couples molecular states. We consider a number of pulse sequences, finding that one can rapidly transfer atoms to a strongly correlated state with nearly unit efficiency. Following a proposal by Popp et al. [13], experimentalists at Stanford have achieved considerable success with a related procedure, where one slowly increases the rate of rotation, adiabatically transferring bosonic atoms from an initially non-rotating state, to an analog of the Laughlin state (S. Chu, private communication). One could also imagine implementing more sophisticated protocols such as rapid adiabatic passage [16].

To achieve sufficient signal to noise, any experimental attempt to study small clusters of atoms must employ an ensemble of identical systems: for example by trapping small numbers of atoms at the nodes of an optical lattice. When formed by sufficiently intense lasers, this lattice will isolate the individual clusters, preventing any "hopping" from one node to another. We will not address the very interesting question of what would happen if the barriers separating the clusters were lowered. By using filtering techniques, one can ensure that the same number of atoms sit at each node (S. Chu, private communication). A rotating deformation of each *microtrap* can be engineered through a number of techniques. For example, if the intensity of the lattice beams forming a triangular lattice are modulated in sequence, then a rotating quadrupolar deformation is be produced. A more versatile technique is to modulate the phases between counterpropagating lattice beams. Changing these phases uniformly translates the lattice sites. If one moves the lattice sites around faster than the characteristic times of atomic motion 10^{-5} s but slow compared to the times for electronic excitations 10^{-15} s then the atoms see a time averaged potential. This technique, which is closely related to the time orbital potential traps pioneered at JILA [17], can produce almost arbitrary time dependent deformations of the individual traps which each of the clusters

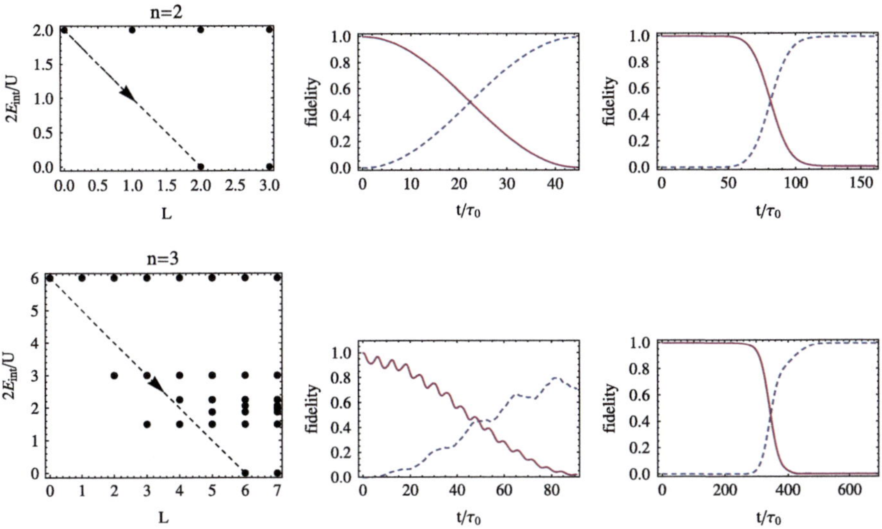

Fig. 9.1 (Color online) Transferring small clusters from non-rotating ground state to $\nu = 1/2$ Laughlin state using rotating quadrupolar ($m = 2$) deformations. *Left*: Interaction energy (in units of $U/2$) of quantum states of harmonically trapped two dimensional clusters as a function of total angular momentum projection L in units of $\hbar$. Excitation paths are shown by *arrows*. *Central*: squared overlap (fidelity) of $|\psi(t)\rangle$ with the initial (*solid*) and final (*dashed*) states as a function of the duration of a square pulse. *Right*: Fidelities as a function of time for an optimized Gaussian pulse of the form $e^{-(t-t_0)^2/\tau^2}$. Time is measured in units of $\tau_0 = \hbar/U \sim 10^{-4}$ s. For $n = 2$, the peak perturbation amplitude is $V_p = 0.05(U/2)$, $\omega - \Omega_p = 2.0(U/2)$, and a Gaussian pulse time of $\tau = 24\tau_0$. For $n = 3$, $\tau = 102\tau_0$ and $\omega - \Omega_p = 2.046(U/2)$ and $2.055(U/2)$ for the Gaussian and square cases, respectively. For $n = 3$, nonlinear effects (coupling with near-resonant levels) shifted the optimal frequency away from the linear response expectation, $\omega - \Omega_p = 2(U/2)$

experiences (S. Chu, private communication). Each cluster feels the same potential.

Once created, the ensemble of clusters can be experimentally studied by a number of means. In situ probes such as photoassociation [18] and RF spectroscopy [19] reveal details about the interparticle correlations. In the regime of interest, time-of-flight expansion, followed by imaging, spatially resolves the ensemble averaged *pre*−expansion density. This result follows from the scaling form of the dynamics of lowest Landau level wavefunctions [14].

We model a single cluster as a small number of two-dimensional harmonically trapped bosonic atoms. The two dimensionality can be ensured by increasing the intensity of the lattice beams in the perpendicular direction. Neglecting the zero-point energy, one finds that in the frame rotating with frequency Ω, the single particle harmonic oscillator eigenstates have the form $E_{jk} = \hbar(\omega - \Omega)k + \hbar(\omega + \Omega)j$; $(j, k = 0, 1, \ldots)$. In typical lattices, the interaction energy $U/\hbar \sim 10$ kHz is small compared to the small oscillation frequency $\omega \sim 100$ kHz [20]. Therefore the many-body state will be made up of single particle states with $j = 0$: the lowest Landau level, with wavefunctions of the form

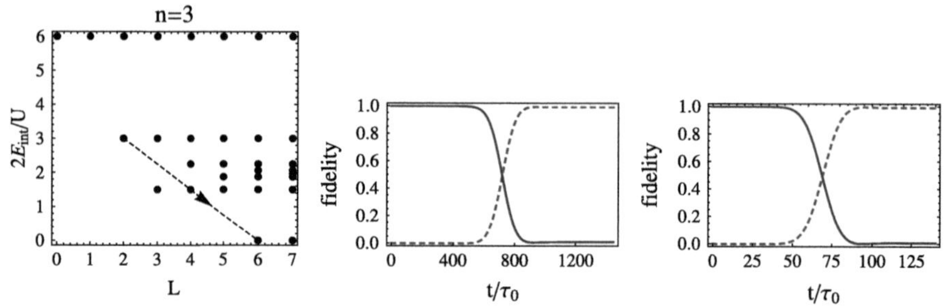

Fig. 9.2 (Color online) Using a rotating m-fold symmetric perturbation to drive $n = 3$ particle clusters from $L = 2$ to the $\nu = 1/2$ Laughlin state. *Left*: path on the energy level diagram. *Center*: second-order process coming from a deformation with $m = 2$. *Right*: direct transition produced with $m = 4$. *Solid (dashed) lines* are fidelities with the initial (Laughlin) state. In both cases the peak deformation is $V_p = 0.05(U/2)$. Both use a Gaussian pulse. The frequencies and pulse times τ we used for $m = 2, 4$ were $\omega - \Omega_p = (3.00/2)(U/2), 3.035(U/2)$ and $\tau/\tau_0 = 218, 21$. *Note* how much more rapid the direct process is

$$\psi_k(x, y) = (\pi k!)^{-1/2} z^k e^{-z^* z/2}, \tag{9.1}$$

where $z = (x + iy)/d$ with $d = \sqrt{\hbar/M\omega}$ is the complex representation of the coordinate in the plane measured in units of the oscillator length, where M is the atomic mass. Including interactions, the many-body Hamiltonian for a single cluster is then

$$H_{LLL} = \sum_j j\hbar(\omega - \Omega) a_j^\dagger a_j + \sum_{jklm} V_{jklm} a_j^\dagger a_k^\dagger a_l a_m \tag{9.2}$$

where a_m is the annihilation operator for the single particle state ψ_m. For point interactions the interaction kernel is

$$V_{jklm} = \frac{U}{2} \delta_{j+k-l-m} 2^{-(j+k)} \frac{(j+k)!}{\sqrt{j!k!l!m!}}, \tag{9.3}$$

where $U = \sqrt{2/\pi}\hbar^2 a/(Md_z d^2)$ is the on-site interaction between two particles in the same well, a is the three dimensional s-wave scattering length and d_z is the oscillator length in the transverse direction. As has been explored in depth by previous authors [1–15], for a given total number of particles n, and angular momentum projection L, the Hamiltonian (9.2) is a finite matrix which is readily diagonalized. Example spectra are shown in Figs. 9.1, 9.2, and 9.3. We plot the spectra as energy versus angular momentum, with $\Omega = \omega$. Spectra at other rotation speeds are readily found by "tilting" the graphs—the energy of a state with angular momentum projection L is simply shifted up by $\hbar(\omega - \Omega)L$.

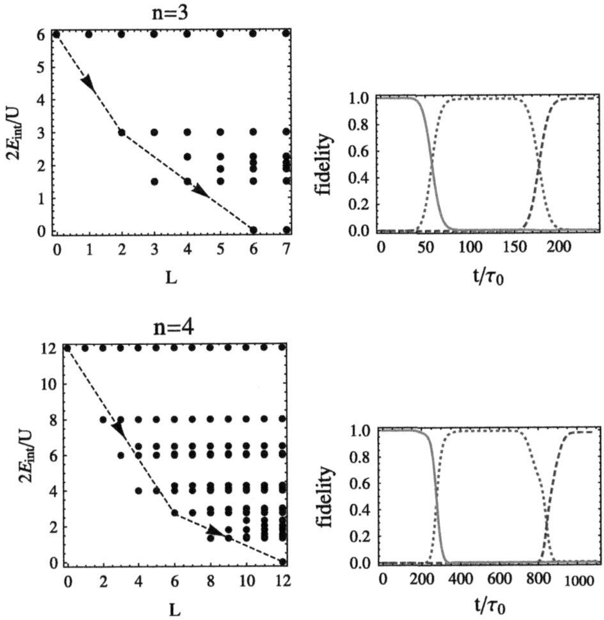

Fig. 9.3 (Color online) Transferring atoms using multiple pulses. *Left*: paths from initial to Laughlin states for $n = 3, 4$. *Right*: *Solid line* is the fidelity with the initial state, *dotted* with the intermediate $(L, E) = (2\hbar, 3(U/2))$ state, and the *dashed line* with the Laughlin state. All pulses are Gaussians. Despite using multiple pulses, this technique is faster than using a higher order $m = 2$ pulse. The frequencies (Ω_p), shape (m), and pulse times (τ) for the $N = 3$ sequence were $\hbar(\omega - \Omega_p)/(U/2) = 3.00, 3.035$, $m = 2, 4$, and $\tau/\tau_0 = 16.95, 19.2$. For both, $V_p = 0.05(U/2)$. For $N = 4$, using two pulses with $m = 2$ and $V_p = 0.2(U/2)$, we achieve $> 98\%$ fidelity after a total two-pulse sequences with $\hbar(\omega - \Omega_p)/(U/2) = 3.130, 1.0376$ and $\tau/\tau_0 = 82.5, 87.0$

We imagine applying to each cluster a rotating single particle potential (in the lab frame) of the form

$$H_S(r,t) = V_p(t)\left[z^m e^{im\Omega_p t} + (z^*)^m e^{-im\Omega_p t}\right], \tag{9.4}$$

where m determines the symmetry of the deformation (e.g. $m = 2$ is a quadrupolar deformation), the envelope function $V_p(t)$ is the time-dependent amplitude of the deformation, and Ω_p is the frequency at which the perturbation rotates. We will mainly focus on the case $m = 2$. When restricted to the lowest Landau level, this potential generates a coupling between the many-body states which in the rotating frame is expressed as

$$H_S = V_p(t) \sum_l v_{lm}(e^{im(\Omega_p - \Omega)t} a^\dagger_{l+m} a_l + \text{H.C.}) \tag{9.5}$$

with $v_{lm} = 2^{-m/2}(l+m)!/\sqrt{l!(l+m)!}$. As such it only couples states whose total angular momentum projection differs by m. For our calculation will work in the co-rotating frame with $\Omega = \Omega_p$, where the only time dependence is given by $V_p(t)$.

We wish to implement a π-pulse, where the amplitude $V_p(t)$ is engineered so that after the pulse, a cluster is transferred from its initial state to a target state of our choosing. If the perturbation coupled only two states, this would be a straightforward procedure. The frequency Ω_p is selected so that the initial and target state are degenerate in the rotating frame. For any finite V_p, the system Rabi flops between the two coupled states, and by turning off the perturbation at the right time one ends up in the target state with unit probability. The present example is more complicated, as there are many states coupled by the perturbation. The basic idea however remains sound: one still chooses Ω_p to make the initial and final state degenerate. The time dependence of $V_p(t)$ should be tailored to minimize the coupling to unwanted states. These stray couplings could be particularly disasterous, because the coupling between the initial and target state are generically quite high order in V_p. As a particularly relevant example, we consider transferring clusters from the ground state (with $L = 0$) to the $\nu = 1/2$ Laughlin state $\psi_L(z_1, \ldots, z_n) = \prod_{i<j}(z_i - z_j)^2 \exp(-\sum_j |z_j|^2/4l_B^2)$, which has angular momentum $n(n-1)$. Using a perturbation with $m = 2$, this requires a $n(n-1)/2$-order process. A picturesque way of thinking about the dynamics in the presence of the perturbation is to map the problem onto the motion of a particle on a complicated "lattice". The states of the unperturbed system are analogous to "lattice sites", while the perturbation produces a "hopping" between sites. The goal is to engineer a time-dependent hopping which efficiently moves the "particle" from a known starting position to a desired ending position. The transfer efficiency is measured by the probability that the system is in the target state ψ_T at the end of the time evolution: we plot this probability—known as the fidelity—as a function of time, given by $f(t) = |\langle\psi_T|\psi(t)\rangle|^2$.

As this analogy emphasizes, the problem of transferring a quantum system from one state to another is generic. Müller et al. [21] recently considered how one can optimize pulse shapes to produce high order Bragg diffraction, while avoiding transferring atoms into unwanted momentum states. These authors developed a formalism for calculating the fidelity by adiabatically eliminating the off-resonant states. They found that Gaussian pulse shapes greatly outperformed simple square pulses. This result is natural, as the smoother pulses have a much smaller bandwidth.

We numerically solve the time dependent Schrödinger equation, truncating our Hilbert space at finite total angular momentum $L = n(n-1) + 4$ for $n = 2, 3$ and $L = n(n-1) + 8$ for $n = 4$. We have numerically verified that changing this cutoff to higher values has negligible effects. Figure 9.1 shows f in the case of $n = 2$ and $n = 3$ for square and Gaussian pulses. For the square pulse the fidelity is shown as a function of pulse length. For the Gaussian pulse, a fixed pulse duration is used, and the fidelity is shown as a function of time. For $n = 2$, where only two states are involved, the pulse shape is irrelevant. For $n = 3$, where there is a near-resonant state with $L = 4$, the Gaussian pulse shape greatly outperforms the square pulse, producing nearly 100% transfer efficiency in 10's of ms, even for a very weak perturbation.

For $n > 3$ we find that these high order processes become inefficient. For $m = 2$ the coupling between the initial and final state scale as $(V_p/U)^{n(n-1)/2}$, making transfer times unrealistical long unless one drives the system into a highly non-linear regime. As illustrated in Fig. 9.2, this difficulty can be mitigated by using perturbations with higher m. There, for illustration, we consider exciting a three-particle cluster from the lowest energy $L = 2$ state to the $L = 6$ Laughlin state. The second order $m = 2$ pulse requires much longer than the first order $m = 4$ pulse. An interesting aside is that one would naively have expected that the resonant $l = 4$ state would make the second-order process extremely inefficient. It turns out that the coupling to that state is fortuitously zero.

Further improved scaling can be arranged by using a sequence of π pulses. One transfers the cluster from one long-lived state to another. Since the number of pulses scales as the angular momentum, the transfer time is then quadratic in the angular momentum, rather than exponential. One can also tailor the path to maximize the fidelity of each step. Some two-pulse sequences are shown in Fig. 9.3. The guiding principle in designing the pulse sequences is that in each step one wants as few as possible near-resonant intermediate states.

9.3 Summary

We have shown it is possible to use time dependent trap perturbations to coherently transfer boson clusters from nonrotating ground states to analogs of fractional quantum hall states. We achieve fidelity $f > 99\%$ for $n = 2, 3$ using very weak rotating $m = 2$ deformations, whose duration is of order tens of ms. Using a two-pulse sequence, we achieve similar results for $n = 4$. We find that smooth Gaussian pulses are much more effective than square pulses, and that further efficiency can be gained by using higher order perturbations of the form z^m with $m > 2$.

We briefly compare our technique with Ref. [13]'s proposal. While our approaches share the use of a rotating time-averaged optical lattice potential, our proposal offers significant differences and advantages. While Ref. [13] suggests an adiabatic evolution, we propose a coherent evolution—analogous to a Rabi oscillation—to the Laughlin state. This has the advantage of being faster, easier to implement, and more robust. For a slightly smaller perturbation relative to the adiabatic method, we achieve fidelity ~ 1 in contrast to the adiabatic method's 0.97 fidelity. Moreover, our method requires half the time. More importantly, the adiabatic method requires carefully navigating a path through possible rotating potential strengths and frequencies as a function of time. In contrast, our method requires only setting the pulse duration and strength, and is thus more easily implementable and less susceptible to small experimental errors.

This technique will allow the efficient creation of bosonic quantum Hall puddles—a state of matter which has not yet been observed. The clusters produced will be orbitally entangled, have strong interparticle correlations, have fractional excitations, and possess topological orders [1–15]. Although current experiments, and the present theory, is focussed on the small-atom limit, it would be exciting to apply these techniques to larger collections of atoms, producing true analogs of fractional quantum Hall states. The main difficulty is that the spectra become dense as n increases, requiring one to set Ω_p to extremely high precision. By carefully choosing the trajectory, taking advantage of gaps in the spectrum, one might be able to overcome such difficulties.

Finally, we mention that our approach allows one to drive the system into almost arbitrary excited states. This may, for example, be important for using quantum hall puddles in a topological quantum computing scheme [22].

Acknowledgements We would like to thank Steven Chu and Nathan Gemelke for discussions about experiments and protocols. This material is based upon work supported by the National Science Foundation through grant No. PHY-0758104.

References

1. N.K. Wilkin, J.M.F. Gunn, R.A. Smith, Phys. Rev. Lett. **80**, 2265 (1998)
2. N.R. Cooper, N.K. Wilkin, Phys. Rev. B **60**, R16279 (1999)
3. S. Viefers, T.H. Hansson, S.M. Reimann, Phys. Rev. A **62**, 053604 (2000)
4. N.K. Wilkin, J.M.F. Gunn, Phys. Rev. Lett. **84**, 69 (2000)
5. N.R. Cooper, N.K. Wilkin, J.M.F. Gunn, Phys. Rev. Lett. **87**, 120405 (2001)
6. B. Paredes, P. Fedichev, J.I. Cirac, P. Zoller, Phys. Rev. Lett. **87**, 010402 (2001)
7. J. Sinova, C.B. Hanna, A.H. MacDonald, Phys. Rev. Lett. **89**, 030403 (2002)
8. S. Viefers, J. Phys. Condens. Matter **20**, 123202 (14pp) (2008)
9. J.W. Reijnders, van F.J.M. Lankvelt, K. Schoutens, N. Read, Phys. Rev. Lett. **89**, 120401 (2002)
10. B. Paredes, P. Zoller, J.I. Cirac, Phys. Rev. A **66**, 033609 (2002)
11. N. Regnault, T. Jolicoeur, Phys. Rev. Lett. **91**, 030402 (2003)
12. U.R. Fischer, P.O. Fedichev, A. Recati, J. Phys. B At. Mol. Opt. Phys. **37**, S301 (2004)
13. M. Popp, B. Paredes, J.I. Cirac, Phys. Rev. A **70**, 053612 (2004)
14. T. Ho, E.J. Mueller, Phys. Rev. Lett. **89**, 050401 (2002)
15. N. Read, N.R. Cooper, Phys. Rev. A **68**, 035601 (2003)
16. K. Bergmann, H. Theuer, B.W. Shore, Rev. Mod. Phys. **70**, 10031025 (1998)
17. W. Petrich, M.H. Anderson, J.R. Ensher, E.A. Cornell, Phys. Rev. Lett. **74**, 3352 (1995)
18. T. Kinoshita, T. Wenger, D.S. Weiss, Phys. Rev. Lett. **95**, 190406 (2005)
19. M.Ö. Oktel, L.S. Levitov, Phys. Rev. Lett. **83**, 69 (1999)
20. D. Jaksch, C. Bruder, J.I. Cirac, C.W. Gardiner, P. Zoller, Phys. Rev. Lett. **81**, 3108–3111 (1998)
21. H. Müller, S.-W. Chiow, Q. Long, S. Herrmann, S. Chu, Phys. Rev. Lett. **100**, 180405 (2008)
22. A.Y. Kitaev, Ann. Phys. **303**, 2 (2003)

Chapter 10
Incorporating Arbitrarily Strong On-Site Correlations into Lattice Models

In Chap. 9 I described a protocol to efficiently generate small clusters of fractional quantum Hall states at sites of a deep optical lattice, in the limit where particles were unable to tunnel between sites on experimentally relevant timescales. One motivation for this chapter is to construct a theory of deep lattices that are nevertheless weak enough to allow particle tunneling between sites (analogous to the Bose-Hubbard limit of bosons in an optical lattice introduced in Chap. 2). I construct a theory—specifically an effective lattice model—that captures the dominant behavior for bosons, and essentially identical techniques lead to a description of fermions. In the bosonic case, I construct an analog of the Gutzwiller mean field theory (introduced in Chap. 2) for this generalized model. I show that the resulting model has a phase diagram topologically the same as the Bose-Hubbard model: there are globally insulating phases with fractional quantum Hall puddles at each site and no coherence between them occupying "Mott lobes" in the chemical potential-lattice depth phase diagram, and a superfluid phase with coherence between the fractional quantum Halls states at each site.

This state is intriguing: for example, the superfluid order parameter $\langle a_j \rangle$ is exactly the nonlocal topological order parameter for the FQH state defined by Girvin and MacDonald [1, 2]. By including higher order corrections to our effective description, the problem is similar to those studied in the community interested in networks of quantum Hall systems (e.g., Ref. [3]).

However, the ideas resulting from these considerations apply more generally than arrays of fractional quantum Hall puddles. Most importantly, our model is able to capture the on-site correlations that occur in many varieties of optical lattice experiments, for example in strongly interacting gases with scattering lengths comparable to the lattice spacing, or for site fillings much greater than

This chapter was adapted from "On-site correlations in optical lattices: Band mixing to coupled quantum Hall puddles" by Kaden R.A. Hazzard and Erich J. Mueller, published in Physical Review A **81**, 031602(R) (2010).

K. R. A. Hazzard, *Quantum Phase Transitions in Cold Atoms and Low Temperature Solids*, Springer Theses, DOI: 10.1007/978-1-4419-8179-0_10,

unity. When more than one particle is put on a site, instead of multiply occupying the lowest band (lowest oscillator state) as assumed in the derivation of Chap. 2, particles partially occupy higher bands to due to interactions. Our model captures arbitrary on-site correlations even in the mean field approximation.

We also estimate model parameters for various situations, and surprisingly find that even for common ^{87}Rb experiments, with $a/d \approx 0.01$, quantitative renormalizations of $\sim 10\%$ manifest for site fillings as low as $n \sim 3$. Since the publication of our paper, some of these have been measured, for example in Ref. [4]

10.1 Chapter Overview

We extend the standard Bose-Hubbard model to capture arbitrarily strong on-site correlations. In addition to being important for quantitatively modeling experiments, for example, with Rubidium atoms, these correlations must be included to describe more exotic situations. Two such examples are when the interactions are made large via a Feshbach resonance, or when each site rotates rapidly, making a coupled array of quantum Hall puddles. Remarkably, even the mean field approximation to our model includes all on-site correlations. We describe how these on-site correlations manifest themselves in the system's global properties: modifying the phase diagram and depleting the condensate.

10.2 Body

Optical lattice systems, where a dilute atomic gas is trapped in a periodic potential formed by interfering laser beams, provide a close connection between solid state systems and atomic physics [5]. The models used to describe these systems generally assume that each lattice site's wavefunction is easily built up from single particle states [6]. Here we argue that this approximation is inappropriate to quantitatively model current experiments, and sometimes fails more drastically, e.g., for resonant bosons. We show how to include arbitrary on-site correlations via a generalized Hubbard model, which can be approached by standard methods. By construction, the mean field approximation to our model captures all on-site correlations, contrasting with prior approaches [7–12].

Our method's key idea is to first consider deep lattices, where lattice sites are isolated and then solve the few-body problem on each site. Next, truncating to this few-body problem's low energy manifold, we calculate how tunneling couples the few-body states on neighboring sites.The resulting theory resembles a Hubbard model, but with number-dependent hopping and interaction parameters. We show that the corrections to the ordinary Bose-Hubbard model captured by this theory are crucial to quantitatively describe current Rubidium experiments. They become even more important when the 3D scattering length a becomes a significant fraction of the size of the Wannier states ℓ, such as in recent experiments on Cesium atoms near a Feshbach resonance [13]. This approach is also essential to describe more exotic on-site correlations; as one example, one can rotate each lattice site, creating a lattice of coupled "quantum Hall puddles" [14, 15]. Related ideas can be applied to double well lattices and coupled "plaquettes" of four sites [16, 17]. We explore the impact of the on-site physics on the extended system's phase diagram.

Our approach is most simply illustrated by a single-component Bose gas in a cubic sinusoidal lattice potential $V_p(x,y,z) = V_0 \sum_{\eta=x,y,z} \sin^2(\pi\eta/d)$ with Hamiltonian

$$H_f = \int d^3r \Big[\psi^\dagger(\mathbf{r}) \Big(-\frac{\hbar^2}{2m}\nabla^2 - \mu + V_p(\mathbf{r}) \Big) \psi(\mathbf{r}) + \frac{2\pi\hbar^2 a}{m} \psi^\dagger(\mathbf{r})\psi^\dagger(\mathbf{r})\psi(\mathbf{r})\psi(\mathbf{r}) \Big], \tag{10.1}$$

where m is the particle mass, μ is the chemical potential, and ψ and $\psi^\dagger$ are bosonic annihilation and creation operators. Adding an additional trapping potential presents no additional difficulties.

Constructing the effective Hamiltonian. For each isolated site, we proceed to build up the many-body states from the solution of the n-body problem at site $j : \langle \mathbf{r_1}, \ldots, \mathbf{r_n} \overline{| |n\rangle}_j = \psi_n(\mathbf{r_1} - \mathbf{R_j}, \ldots, \mathbf{r_n} - \mathbf{R_j})$, which obeys $H_j \overline{|n\rangle}_j = \epsilon_n \overline{|n\rangle}_j$ where H_j is the same as Eq. (10.1)'s H_f, except replacing the periodic potential V_p there with an on-site potential V_j. A convenient approximation is to take $V_j(\mathbf{r}) = (m\omega^2/2)(\mathbf{r} - \mathbf{R_j})^2$ with $\omega = 2\sqrt{V_0 E_R}/\hbar$, the harmonic approximation to the site located at $\mathbf{R_j}$. For each site filling n, we restrict our on-site basis to the lowest energy n-body state; however, including a finite number of excited states is straightforward. Note that even in the non-interacting case these states are *not* Wannier states. The principle difference is that states defined in this way are non-orthogonal. From these, however, one can construct a new set of orthogonal states $|n\rangle_j$, which hold similar physical meaning. In the noninteracting limit, the $|n\rangle_j$ approximate the Wannier states.

Because the single-site wavefunctions decay like Gaussians, it typically suffices to build up the effective Hamiltonian from neighboring sites. In particular, consider two sites L and R , and the space spanned by $\overline{|n_L, n_R\rangle} = \overline{|n_L\rangle}_L \otimes \overline{|n_R\rangle}_R$, with overlaps $S^{(mn)} = \overline{\langle m, n| |m+1, n-1\rangle}$. To lowest order in the overlaps, we can define orthogonal $|n_L, n_R\rangle$ by taking $|n_L, n_R\rangle = \overline{|n_L, n_R\rangle} - (1/2) [S^{(n_R,n_L)} \overline{|n_R+1, n_L-1} + S^{(n_L,n_R)} \overline{|n_R-1, n_L+1\rangle}]$.

Within this restricted basis, the effective Hamiltonian for these two sites is $H_{\text{eff}} = \sum_{n,m,n',m'} |n', m'\rangle \langle n', m'| H_f |n, m\rangle \langle n, m|$. Evaluation to lowest order in $S^{(mn)}$

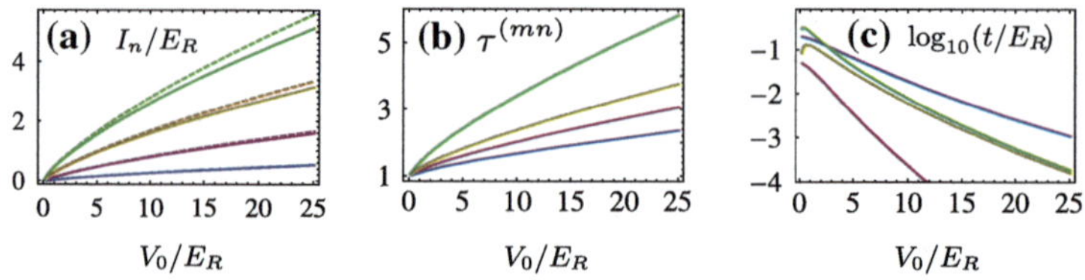

Fig. 10.1 (*Color online*) **a** On-site energy with non-interacting energies subtracted off, $I_n = E_n - (3\hbar\omega/2 - \mu)n$, and energies scaled by $E_R = \hbar^2\pi^2/(2md^2)$, using typical ^{87}Rb parameters: lattice spacing $d = 532$ nm, scattering length $a = 5.32$ nm. *Dashed*: neglecting on-site correlations, *solid*: including correlations. *Bottom* to *top curve*: $n = 2, 3, 4, 5$. **b** Representative hopping matrix elements with on-site correlations, relative to those neglecting on-site correlations, $\tau^{(mn)} \equiv t^{(mn)}/(t\sqrt{(m+1)n})$, as a function of lattice depth V_0 on a scale. *Bottom* to *top line*: $t^{(03)}, t^{(31)}, t^{(05)}, t^{(35)}$. **c** For comparison, $t^{(01)}/E_R$ calculated from the exact Wannier states (*upper curve*) along with our Gaussian approximation to it with and without non-orthogonality corrections (second and third highest, respectively); also shown is the next-nearest neighbor hopping matrix element (*bottom curve*). The effective Hamiltonian parameters are calculated perturbatively in a/d for a Gaussian ansatz

yields on-site energy terms $\sum_{n,m}(E_n + E_m)|n,m\rangle\langle n,m|$ and a "hopping" term $-\sum_{nm} t^{(mn)}|m+1, n-1\rangle\langle m,n| + \text{H.c.}$ with

$$\begin{aligned} E_n &= \overline{\langle n|}H_f\overline{|n\rangle}, \\ t^{(mn)} &= -\overline{\langle m+1, n-1|}H_f\overline{|m,n\rangle} + \frac{S^{(mn)}}{2}(E_m + E_n). \end{aligned} \tag{10.2}$$

Additionally there is an interaction term $U = \sum_{nm}[U_{LL}^{(n)} + U_{RR}^{(m)} + U_{LR}^{(n,m)}]|m,n\rangle\langle m,n|$ with $U_{LL}^{(m)} = U_{RR}^{(m)} = E_m$ and

$$U_{LR}^{(n,m)} = \overline{\langle m,n|}H_f\overline{|m,n\rangle} - E_m - E_n \tag{10.3}$$

to $O(S^2)$, consistent with the rest of our calculations. In the remainder of this paper we will neglect the off-site interaction, Eq. 10.3, and the last term in Eq. 10.2. The former is rigorously justified as it falls of exponentially faster than the other interaction terms. Formally,the non-orthogonality contribution to Eq. 10.2 is suppressed only by a factor of $(V_0/E_R)^{1/4}$ with $E_R = \hbar^2\pi^2/(2md^2)$, but as shown in Fig. 10.1, it is typically small.

The simplest many-site Hamiltonian which reduces to this one in the limit of two sites is

$$\begin{aligned} H = &-\sum_{\langle i,j\rangle;m,n} t_{ij}^{(mn)}|m+1\rangle_i|n-1\rangle_j\langle m|_i\langle n|_j \\ &+\sum_{i,n} E_n|n\rangle_i\langle n|_i, \end{aligned} \tag{10.4}$$

where $\sum_{\langle i,j\rangle}$ indicates a sum over nearest neighbors i and j. At higher order, one generates more terms such as next nearest neighbor hoppings, pair hoppings, and longer range interactions.

Calculating the Hamiltonian parameters. Here we consider the cases of weak interactions, resonant interactions, and coupled quantum Hall puddles.

In the limit of weak interactions, one can estimate the parameters in Eq. 10.4 by taking the on-site wavefunction to be $\psi_n \propto \exp(-\sum_{j=1}^{n} r_j^2/2\sigma_n^2)$, with variational width σ_n. To leading order in a/d we find $E_n = E_R[(3\sqrt{V_0/E_R} - \mu/E_R)n + (U/2)$ $n(n-1)\left(1 - \frac{3\pi}{4\sqrt{2\pi}}(a/d)(n-1)(V_0/E_R)^{1/4}\right)]$ with $U = (a/d)\sqrt{2\pi}(V_0/E_R)^{3/4}$ and $t^{(mn)} = t\sqrt{n(m+1)}[1 + \frac{\sqrt{2}a\pi^{5/2}}{4d}(m+n-1)(V_0/E_R)^{3/4}]$ with $t = V_0(\pi^2/4 - 1)$ $e^{-(\pi^2/4)\sqrt{V_0/E_R}}$. Note that as expected, interaction spreads out the Wannier functions, increasing the $t^{(mn)}$'s and decreasing the E_n's. Figure 10.1 shows several of the resulting $t^{(mn)}$ as a function of V_0 for parameters in typical optical lattice experiments with ^{87}Rb. Also shown are $t^{(01)}$ and the next-nearest-neighbor hopping, t_{nnn}, calculated from the exact Wannier states. Our estimates are consistent with previous work regarding t's n-dependence [7–9], validating our approach. As can be seen in Fig. 10.1c, the relative size of the next nearest neighbor hopping t_{nnn}/t is 10% (1%) for $V_0 = 3E_R$ ($V_0 = 10E_R$), justifying our approximation of including only nearest neighbor overlaps to describe the system near the Mott state. Figure 10.1 also illustrates that the Gaussian approximation only qualitatively captures the behavior even for non-interacting particles. *We also see that even for this weakly interacting case the number dependence of t is crucial for a quantitative description of the experiments.* Similarly, the number dependence of the on-site interaction is quantitatively significant. This latter deficiency of the standard Hubbard model has been noted in the past, for example by the MIT experimental group [18].

For more general experimental systems, one needs to include still more on-site correlations. As our first example, we consider lattice bosons near a Feshbach resonance [19, 20], describing, for example, ongoing Cesium atom experiments [13].

We restrict ourselves to site occupations $n = 0, 1, 2$, for which we have exact analytic solutions to the on-site problem for arbitrary a in terms of confluent hypergeometric functions [21]. Figure 10.2 shows graphs of E_n and $t^{(mn)}$ rescaled by $\hbar\omega$ as a function of a in the deep lattice limit.

As Fig. 10.2b illustrates dramatically, the hopping from and to doubly-occupied sites is strongly suppressed near the Feshbach resonance when atoms occupy the lowest branch, and is enhanced for the next-lowest branch. The former has implications for studies of boson pairing on a lattice [19, 20], showing that one must dramatically modify previous models near resonance, and, as will be discussed more below, the latter implies a substantial reduction of the $n = 2$ Mott lobe's size for repulsive bosons.

We give one further example, namely the case, similar to the one discussed in [14, 15], where the individual sites of the optical lattice are elliptically deformed

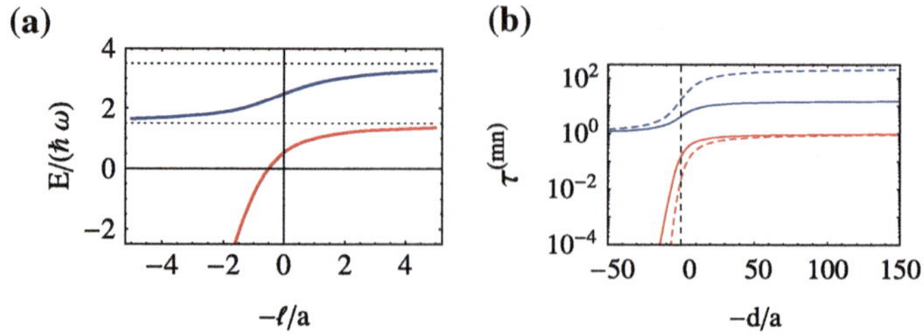

Fig. 10.2 (*Color online*) *Left*: On-site two-particle energy as a function of scattering length a rescaled by the on-site harmonic oscillator energy $\hbar\omega = 2\sqrt{V_0 E_R}$, for the two lowest energy branches. The corresponding characteristic length is $\ell = \sqrt{\hbar/m\omega}$. *Right*: Log plot of rescaled hopping matrix elements $\tau^{(mn)} \equiv t^{(mn)}/(t\sqrt{(m+1)n})$. Solid and dashed curves are $t^{(11)}/(\sqrt{2}t)$ and $t^{(12)}/(2t)$, respectively. We have chosen the lattice depth $V_0 = 15E_R$; this affects only the horizontal scale. In the ordinary Bose-Hubbard model, $t^{(mn)}/(\sqrt{(m+1)n}t) = 1$ for all m, n, as confirmed by this figure's $a = 0^-$ lowest (*red*) branch and $a = 0^+$ second (*blue*) branch limits. The resonance at $-d/a = 0$ separates the molecular side (*left*), from the atomic side (*right*). Note that $t^{(10)}/t$ (not shown) is universally equal to unity regardless of interaction strength, since interatomic correlations are absent when there is a single particle per site

and rotated about their center. This is accomplished by rapidly modulating the phase of the optical lattice lasers to generate an appropriate time-averaged optical potential. At an appropriate rotation speed Ω the lowest energy n-particle state on each site is a $\nu = 2$ Laughlin state $\psi_n(\mathbf{r_1}, \ldots, \mathbf{r_n}) = \mathcal{N}_\mathbf{n}\left[\prod_{\mathbf{i<j=1}}^{\mathbf{n}}(\mathbf{w_i} - \mathbf{w_j})^\mathbf{2}\right]$ $e^{-\sum_j |w_j|^2/(4\ell^2)}$, where we define $w_j \equiv x_j + iy_j$, and $\mathcal{N}_n$ is a normalization factor with phase chosen to gauge away phase factors appearing in $t^{(mn)}$. Truncating to this set of states for $n = 0, 1, 2$, we produce an effective Hubbard model of the same form as Eq. 10.4. The hopping parameters for asymptotically deep lattices $V_0/E_R \gg 1$ are $t^{(01)} = t$, $t^{(02)} = t(\pi^2/32)(V_0/E_R)^{1/2}$, and $t^{(12)} = t(\pi^4/1024)$ (V_0/E_R) where $E_R = \hbar^2\pi^2/(2md^2)$ is the recoil energy, and t is the same as in the weakly interacting case treated above; the interaction parameters are $E_m = \frac{\omega-\Omega}{2} m(m-1) - \mu m$. One particularly interesting aspect of this model of coupled quantum Hall puddles is that when the system is superfluid, the order parameter is exactly the quantity defined by Girvin and MacDonald [1, 2] to describe the nonlocal order of a fractional quantum Hall state. Thus when one probes the superfluid phase stiffness, one directly couples to this quantity.

Mean-field theory. The true strength of our approach is that the resulting generalized Hubbard model is amenable to all of the analysis used to study the standard Bose-Hubbard model. In particular, we can gain insight from a Gutzwiller mean-field theory (GMFT) [6, 22]. This approximation to the ordinary Bose-Hubbard model gives moderate quantitative agreement with more sophisticated methods: for example, the unity site filling MI/SF transition on a 3D cubic lattice occurs at $(t/U)_c = 0.03408(2)$ while GMFT yields $(t/U)_c = 0.029$ [23].

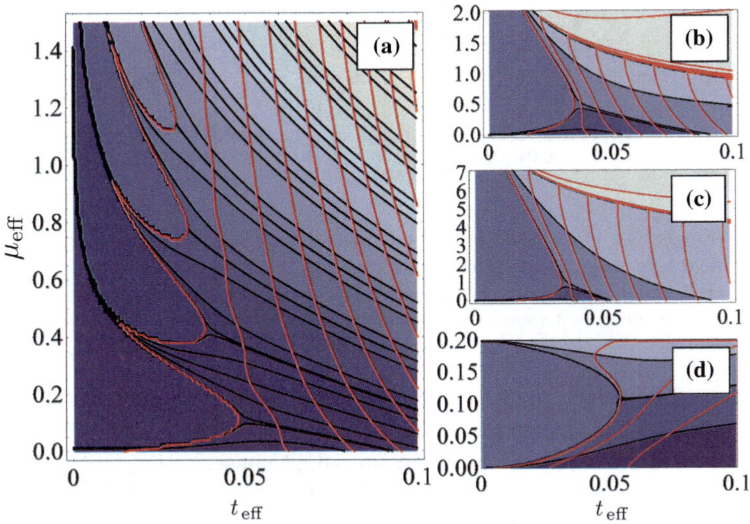

Fig. 10.3 (*Color online*) Representative Gutzwiller mean-field theory phase diagrams, showing constant density (*black*, roughly horizontal) and constant $\xi \equiv \zeta_1 + \zeta_2 + \zeta_3$ (*red*, roughly vertical) contours: ξ, similar to the condensate density, is a combination of the mean fields ζ_m, defined after Eq. 10.2. Density contours are $n = \{0.01, 0.2, 0.5, 0.8, 0.99, 1.01, 1.2, \ldots\}$ and order parameter contours are $\xi = \{0.2, 0.4, \ldots\}$, except (**d**) where we take contours $\xi = \{0.02, 0.04, \ldots\}$. The phase diagrams are functions of $\mu_{\rm eff} \equiv \mu/E_R$ and $t_{\rm eff} \equiv \exp\left(-\sqrt{V_0/E_R}\right)$, where the lattice depth V_0 is the natural experimental control parameter. We plot versus $t_{\rm eff}$, instead of V_0, as this is closer to the Hamiltonian matrix elements and more analogous to traditional visualizations of the Bose-Hubbard phase diagram. **a** Ordinary Bose-Hubbard model for $a = 0.01d$, **b** lattice bosons restricted to fillings $n = 0, 1, 2$ with $a = 0.01d$, on the next to lowest energy branch on the $a > 0$ side of resonance, **c** lattice boson model with $a = d$, and **d** FQH puddle array model taking $\omega - \Omega = 0.1E_R$ (see text for details). Parts (b, c) use the Hamiltonian parameters from the exact two-particle harmonic well solution

In the ground state $|\Psi\rangle$, we introduce mean fields $\xi_m \equiv \langle \Psi m + 1\rangle\langle m|\Psi\rangle$. Neglecting terms which are quadratic in $\delta L^i_m = |i, m+1\rangle\langle i, m| - \xi_m$, the Hamiltonian is $H_{MF} = \sum_i H_{MF,i}$ with

$$H_{MF,i} = E_{n_i}|n\rangle_i\langle n|_i - z\sum_m \left[\zeta_m|m-1\rangle_i\langle m|_i - \zeta_m\xi_{m-1} + \text{H.c.}\right], \qquad (10.5)$$

where H.c. denotes Hermitian conjugate, z is the lattice coordination number, and $\zeta_n = \sum_m \xi_m t^{(mn)}$.

Truncating the number of atoms on a site to $n \le n_{\max}$, we self-consistently solve Eq. 10.5 by an iterative method. We start with trial mean-fields, calculate the lowest energy eigenvector of the $(n_{\max}+1) \times (n_{\max}+1)$ mean field Hamiltonian matrix, then update the mean-fields. We find that it typically suffices to take $n_{\max}$ roughly three times the mean occupation of the sites. Figure 10.3 illustrates how

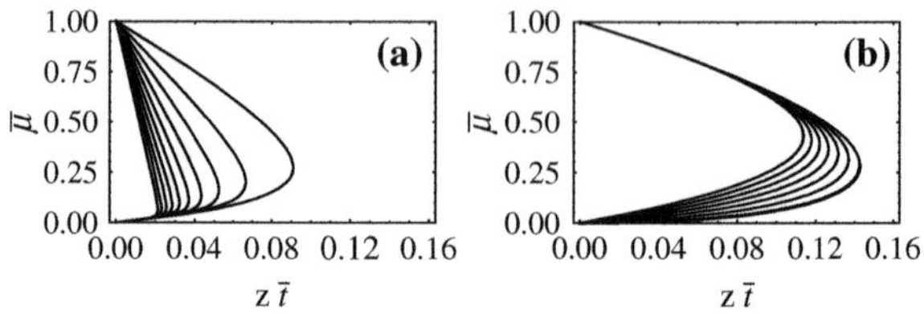

Fig. 10.4 Single lobe of the Mott insulator/superfluid boundary. Complete characterization of the mean-field lobe shape in the $z\bar{t}$–$\bar{\mu}$ plane for all possible $t^{\pm}$'s (see Eq. 10.7 for definitions). **a** Fix $t^{-} = 0.5$, vary t^{+} from 1 (*outer curve*) to 21 (*inner curve*) in steps of 2. **b** Fix $t^{+} = 1.5$, vary t^{-} from 0 (*outer curve*) to 1 (*inner curve*) in steps of 0.15

the density dependence of the parameters introduced by the on-site correlations modify the GMFT phase diagram—particularly the phase boundary's shape, and the density and order parameter in the superfluid phase.

As one would expect, the topology of the MI/SF phase boundaries are similar to that of the standard Bose-Hubbard model, but the Mott lobes' shapes can be significantly distorted. Within mean-field theory the boundary's shape can be determined analytically by taking $|\Psi = \epsilon'|n-1\rangle + \sqrt{1-\epsilon'2-\epsilon^2 f_{n-1}}|n\rangle + \epsilon|n+1\rangle$, and expanding $\langle\Psi|H_{MF}|\Psi\rangle$ to quadratic order in ϵ and ϵ'. The Mott boundary corresponds to when the energy expectation value's Hessian changes sign; this boundary occurs when

$$\begin{aligned}(E_{n+1} - E_n + 2zt^{(n,n+1)})\\ \times (E_{n-1} - E_n + 2zt^{(n-1,n)}) = (2zt^{(n,n)})^2.\end{aligned} \quad (10.6)$$

The five scaled parameters

$$\begin{aligned}\bar{\mu} &\equiv \frac{E_n - E_{n-1}}{E_n}, \quad x_U \equiv \frac{E_{n+1} + E_{n-1} - 2E_n}{E_n},\\ \bar{t} &\equiv \frac{t^{(n,n)}}{E_n}, \quad t^{+} \equiv \frac{t^{(n,n+1)}}{t^{(n,n)}}, \quad t^{-} \equiv \frac{t^{(n-1,n)}}{t^{(n,n)}},\end{aligned} \quad (10.7)$$

completely characterize the shape of a filling-n Mott phase boundary. Varying $\bar{t}$ and $\bar{\mu}$ while fixing the other parameters then maps out a Mott-lobe like feature in the $\bar{t}$ and $\bar{\mu}$ plane, is illustrated in Fig. 10.4.

Summary and discussion. We have demonstrated a novel approach to strongly correlated lattice boson problems. One constructs a model by truncating the on-site Hilbert space to a single state for each site filling n and includes only nearest-site, single particle hoppings. While this approximation captures many multi-band effects, it is not a multi-band Hubbard model and in particular retains the ordinary Bose-Hubbard model's simplicity. In this method, arbitrary on-site correlations may be treated, even in the mean field theory, and consequently it captures the condensate depletion, modified excitation spectra, altered condensate wavefunction, and altered equation of state characteristic of strongly interacting bosons. We have calculated the mean field Mott insulator/superfluid phase boundary analytically and observables across the phase diagram numerically.

Finally, although we have truncated to a single many-body state for each filling n, no difficulty arises from including on-site many-body excitations in the Hamiltonian. These are especially important, for example, for double well lattices and spinor bosons. The ideas also extend straightforwardly to fermions; see Refs. [24–28] for related considerations.

Acknowledgments This material is based upon work supported by the National Science Foundation through grant No. PHY-0758104. We thank Stefan Baur, John Shumway, and Mukund Vengalattore for useful conversations.

References

1. S.M. Girvin, A.H. MacDonald, Phys. Rev. Lett. **58**, 1252 (1987)
2. N. Read, Phys. Rev. Lett. **62**, 8689 (1989)
3. E. Shimshoni, A. Auerbach, A. Kapitulnik, Phys. Rev. Lett. **80**, 3352 (1998)
4. S. Will, T. Best, U. Schneider, L. Hackermüler, D. Lühmann, I. Bloch, Nature **465**, 197 (2010)
5. I. Bloch, J. Dalibard, W. Zwerger, Rev. Mod. Phys. **80**, 885 (2008)
6. D. Jaksch, C. Bruder, J.I. Cirac, C.W. Gardiner, P. Zoller, Phys. Rev. Lett. **81**, 3108 (1998)
7. G. Mazzarella, S.M. Giampaolo, F. Illuminati, Phys. Rev. A **73**, 013625 (2006)
8. A. Smerzi, A. Trombettoni, Phys. Rev. A **68**, 023613 (2003)
9. L. Li, J. Checkelsky, Y. Hor, C. Uher, A. Hebard, R. Cava, N. Ong, Science **321**, 547 (2008)
10. J. Larson, A. Collin, J. Martikainen, Phys. Rev. A **79**, 033603 (2009)
11. O.E. Alon, A.I. Streltsov, L.S. Cederbaum, Phys. Rev. Lett. **95**, 030405 (2005)
12. P.R. Johnson, E. Tiesinga, J.V. Porto, C.J. Williams, New J. Phys. **11**, 093022 (2009)
13. C. Chin, private communication.
14. M. Popp, B. Paredes, J.I. Cirac, Phys. Rev. A **70**, 053612 (2004)
15. S.K. Baur, K.R.A. Hazzard, E.J. Mueller, Phys. Rev. A **78**, 061608 (2008)
16. P. Barmettler, A.M. Rey, E. Demler, M.D. Lukin, I. Bloch, V. Gritsev, Phys. Rev. A **78**, 012330 (2008)
17. L. Jiang, A.M. Rey, O. Romero-Isart, J.J. Garcá-Ripoll, A. Sanpera, M.D. Lukin, Phys. Rev. A **79**, 022309 (2009)
18. G.K. Campbell, J. Mun, M. Boyd, P. Medley, A.E. Leanhardt, L.G. Marcassa, D.E. Pritchard, W. Ketterle, Science **313**, 649 (2006)
19. A.J. Daley, J.M. Taylor, S. Diehl, M. Baranov, P. Zoller, Phys. Rev. Lett. **102**, 040402 (2009)
20. D.B.M. Dickerscheid, U.A. Khawaja, van D. Oosten, H.T.C. Stoof, Phys. Rev. A **71**, 043604 (2005)
21. T. Busch, B. Englert, K. Rzazewski, M. Wilkens, Found. Phys. **28**, 549 (1998)
22. M.P.A. Fisher, P.B. Weichman, G. Grinstein, D.S. Fisher, Phys. Rev. B **40**, 546570 (1989)
23. B. Capogrosso-Sansone, N.V. Prokof'ev, B.V. Svistunov, Phys. Rev. B **75**, 134302 (2007)
24. R.B. Diener, T. Ho, Phys. Rev. Lett. **96**, 010402 (2006)
25. H. Zhai, T. Ho, Phys. Rev. Lett. **99**, 100402 (2007)
26. M. Köhl, H. Moritz, T. Stöferle, K. Günter, T. Esslinger, Phys. Rev. Lett. **94**, 080403 (2005)
27. L. Duan, Phys. Rev. Lett. **95**, 243202 (2005)
28. C. Mathy, D.A. Huse, Phys. Rev. A **79**, 063412 (2009)

Chapter 11
Quantitative Calculation of Parameters for a Model Sufficiently General to Capture all On-Site Correlations

11.1 Background

Chapter 10 introduced a generalized Bose-Hubbard model that is, under appropriate circumstances, capable of describing experimentally important situations, and which is much more general than the ordinary Bose-Hubbard model. However, while we could deduce the *form* of the Hamiltonian, quantitatively calculating the model's parameters is non-trivial.

This chapter develops a method to calculate these model parameters for bosons with scattering length much less than the interparticle spacing (as is relevant to ^{87}Rb) and which is extensible to somewhat stronger interacting systems. It proceeds by numerical path integral Monte Carlo solution of the two (or few) site continuum problem, from which one can extract the relevant lattice description parameters for the many-site problem. This much simpler problem may then be examined with analytic techniques, or by numerical simulation much more efficiently than direct simulation of the many-site continuum problem.

11.2 Introduction, Notation, and Set Up

The Hamiltonian of the Bose-Hubbard model extended to capture on-site correlations is (recalling Eq. 10.4)

$$H = -\sum_{\langle i,j\rangle;m,n} t_{ij}^{(mn)} \big|m+1\rangle_i \big|n-1\rangle_j \langle m\big|_i \langle n\big|_j + \sum_{i,n} \epsilon_n \big|n\rangle_i \langle n\big|_i. \tag{11.1}$$

K. R. A. Hazzard, *Quantum Phase Transitions in Cold Atoms and Low Temperature Solids*, Springer Theses, DOI: 10.1007/978-1-4419-8179-0_11,

There are additional pair (and more) hopping terms and an off-site interaction that we argued are irrelevant in the asymptotically deep lattice. Although I neglect these here, our approach is straightforwardly extended to calculate these parameters.

The Hamiltonian parameters are found by requiring $\langle\psi'|H|\psi\rangle = \langle\psi'|H_f|\psi\rangle$ for all $|\psi\rangle$ and $|\psi'\rangle$ for which $\langle\psi'|H_f|\psi\rangle$ is non-negligible, where H_f is the true continuum Hamiltonian. This yields

$$E_n = \langle n|H_f|n\rangle \tag{11.2}$$

$$t^{(mn)} = -\langle m+1, n-1|H_f|m, n\rangle. \tag{11.3}$$

We work with cold atomic systems with short range potentials, which are described by

$$H_f = \int d^3r \left[\psi^\dagger(\mathbf{r})\left(-\frac{\hbar^2}{2m}\nabla^2 - \mu + V_p(\mathbf{r})\right)\psi(\mathbf{r}) + \frac{2\pi\hbar^2 a}{m}\psi^\dagger(\mathbf{r})\psi^\dagger(\mathbf{r})\psi(\mathbf{r})\psi(\mathbf{r})\right], \tag{11.4}$$

in the limit of interest, where m is the particle mass, μ is the chemical potential, and ψ and $\psi^\dagger$ are bosonic annihilation and creation operators.

11.3 Quantitative Estimates of the Hamiltonian Parameters with Quantum Monte Carlo

11.3.1 General Idea

Although Eq. 10.2 gives the Hamiltonian parameters, there are two fundamental deficiencies of Chap. 10's approach to calculating the model parameters (there, only estimates were attempted). The first deficiency is a technical one: since the few-body problem is generically intractable, we were unable even to evaluate the wavefunctions $|m\rangle$ except in very special circumstances; and evaluating the matrix elements is still more difficult. The second deficiency is more fundamental: even if we could evaluate the $|m\rangle$—the solutions in a harmonic trap—these fail to give exact solutions for the energy dispersion, even in the deep lattice limit, and even for non-interacting particles. Put another way, even if one could in principle calculate the matrix elements exactly, these would not yield the correct energy dispersion. The reason for this is our solutions' incorrect behavior in the "tails" between sites, our Hilbert space does not fully include states with low energy. Formally, we may still truncate to the basis we have, but the zero'th order condition of equating matrix elements of our generalized Bose-Hubbard Hamiltonian with the full Hamiltonian's matrix elements would need to be supplemented with higher order terms in off-diagonal degenerate perturbation theory (e.g. the virtual transitions to states out of this basis).

Thus, although our earlier paper established that it is necessary to go beyond the ordinary Bose-Hubbard model by including on-site correlations with our model, we were unable to quantitatively calculate the correct Hamiltonian parameters. To overcome these difficulties we suggest the use of a quantum Monte Carlo procedure for two sites.

The central idea is that we need the GBHM's parameters only for a two-site system, from which we can construct the model for an arbitrary lattice. To determine these, we should choose our GBHM's parameters to give rise to the same observables as the actual physical system. A first thought would be to match the moments $\langle n_L \rangle, \langle n_L^2 \rangle, \langle n_L^3 \rangle, \ldots$ where n_L is the number operator on the left site (L). However, this provides only a few numbers, each of which is largely redundant with the others, to determine the parameters. It seems that a better way is to match $\langle P_m(\tau) P_n(0) \rangle$ with the operator P_m defined by $P_m \equiv |m, N-m\rangle\langle m, N-m|$; we are working within an N total particle basis and $|m, n\rangle \equiv |m\rangle_L |n\rangle_R$. This approach gives us considerably more information to match—an entire matrix of variables, each of which is a function of imaginary time—yet the problem is still tractable both for the full Hamiltonian and the GBHM (with essentially no more difficulty than calculating the moments). This procedure determines the $t^{(mn)}$'s. The E_m's are determined from the still-simpler one-site problem.

11.3.2 Relating Response to Correlation Functions

To make our idea in the previous section precise, we note that in directly sampling the imaginary time worldlines, path integral Monte Carlo (PIMC) [1] already samples the necessary information to obtain the imaginary time-ordered Green's function

$$\mathcal{G}_{mp}(\tau) = \langle \mathcal{T} P_m(\tau) P_p(0) \rangle \tag{11.5}$$

with the expectation value in the thermal state of the full Hamiltonian, where $\mathcal{T}$ indicates time-ordering. We will concern ourselves with the temperature $T = 0$ limit for the rest of these notes, in which case the average is carried out in the ground state, but this condition can be relaxed if it is helpful to compare with PIMC results.

11.3.3 Implementation of Correlation Function in PIMC

Path integral Monte Carlo naturally measures time-ordered correlation functions. However, we want $\mathcal{G}_{mp}(\tau) = \langle \mathcal{T} P_m(\tau) P_p(0) \rangle$, which would require implementing the many-body wavefunctions $|m\rangle$ and obtaining their evolution and correlations. An approximation allows a much simpler calculation with little loss of accuracy.

The idea is to partition space into two regions; we create an imaginary dividing plane between the left and right wells. We define S_m as the operator that is one is m particles are to the left of the dividing plane. We can then calculate $\langle \mathcal{T} S_m(\tau) S_p(0)\rangle$ by counting at each time slice in the PIMC simulation if p atoms are in the left well at time $\tau = 0$ and m are in the left well at time τ.

We argue that, in the limit where the GBHM is expected to be a good approximation, it is equally accurate to take

$$\langle \mathcal{T} P_m(\tau) P_p(0)\rangle \approx \langle \mathcal{T} S_m(\tau) S_p(0)\rangle. \tag{11.6}$$

This is most concrete in the case of a single particle: the criterion is that of the single band limit, in which case the difference between projection onto the plane on the right and the state on the right comes only in the tails of the right site's Wannier function. Thus, the error in our approximation is of relative order S where S is the overlap between sites, and is on the order of one percent for our systems.

11.4 Solutions for Various Values of $t^{(mn)}$ and E_m

To anchor our intuition, will we do simple limits—non-interacting and non-hopping. Then we turn to numerically solving the general case.

11.4.1 Non-Interacting Case

11.4.1.1 Eigenstates and Eigenvalues

Here I solve the non-interacting case, by which I mean I mean the non-interacting case of the original Hamiltonian, so that one obtains an ordinary Bose-Hubbard model. This is a stronger requirement than just having E_m linear in the GBHM, which could also be called non-interacting.

In the non-interacting case, we can reduce the physics to the single particle physics. For a single particle, there are two energy eigenstates, the symmetric (bonding) and antisymmetric (anti-bonding) orbitals; this is required already by symmetry since the potential is reflection symmetric.

Then, the normalized single particle states are $|\Psi_1^{(\pm)}\rangle' = (1/\sqrt{2})(|0\rangle \pm |1\rangle)$. I use the notation $|m\rangle \equiv |m, N-m\rangle$. This is rewritten in terms of annihilation operators for the left and right sites, L and R, respectively, and the associated creation operators as

$$|\Psi_1^{(\pm)}\rangle' = S_\pm^\dagger|\text{vac}\rangle = \frac{L^\dagger \pm R^\dagger}{\sqrt{2}}|\text{vac}\rangle, \tag{11.7}$$

defining $S_\pm \equiv (1/\sqrt{2})(L \pm R)$. The $|\rangle'$ notation indicates that we are dealing here with the one particle energy eigenstates rather than number eigenstates.

The advantage of writing things this way is the ease with which we build up the many body states; for example, the normalized ground state is simply

$$|\Psi_N\rangle' = \frac{\left(S^\dagger\right)^N}{\sqrt{N!}}|\text{vac}\rangle. \tag{11.8}$$

More generally, the normalized s'th excited state of the N body system, which we denote $\overline{|s\rangle}$ is

$$\overline{|s\rangle} = \frac{1}{\sqrt{s!(N-s)!}}\left(S_+^\dagger\right)^{N-s}\left(S_-^\dagger\right)^s|\text{vac}\rangle \tag{11.9}$$

which puts s particles in the single particle excited state and $N - s$ in the single particle ground state. It immediately follows that the energies defined by $H\overline{|s\rangle} = E_s\overline{|s\rangle}$ are

$$\begin{aligned} E_s &= s\frac{\Delta}{2} - (N-s)\frac{\Delta}{2} \\ &= (s - N/2)\Delta \end{aligned} \tag{11.10}$$

where I defined Δ as the energy splitting between the single particle energy eigenstates, and put the zero of energy halfway between them. In reference to the usual two site non-interacting Bose-Hubbard Hamiltonian, $H_{\text{ni}} = -t(L^\dagger R + R^\dagger L)$, we obtain $\Delta = t$, and ignore the irrelevant constant energy shift. I shall work in terms of Δ in what remains.

11.4.1.2 Expectations of $P_m = |m\rangle\langle m|$

One of the most basic things in which we are interested is the thermal expectation value $\langle P_m\rangle$ with $P_m \equiv |m\rangle\langle m|$. Combinatoric arguments for placing distinguishable particles should give this to be a normalized binomial distribution in the ground state. In a general thermal state we obtain

$$\langle P_m\rangle = \frac{\sum_s e^{-\beta E_s}\overline{\langle s|}P_m\overline{|s\rangle}}{\sum_s e^{-\beta E_s}}. \tag{11.11}$$

We know E_s so to calculate this, we just need to calculate $\overline{\langle s|}P_m\overline{|s\rangle}$. Start by writing $\overline{|s\rangle}$ in the number basis, starting from Eq. 11.9:

$$
\begin{aligned}
\overline{|s\rangle} &= \frac{1}{\sqrt{2^N s!(N-s)!}}(L+R)^{N-s}(L-R)^s|\text{vac}\rangle \\
&= \frac{1}{\sqrt{2^N s!(N-s)!}}\sum_{m_1=0}^{s}\sum_{m_2=0}^{N-s}(-1)^{m_1}\binom{s}{m_1}\binom{N-s}{m_2} \\
&\quad\times (L^\dagger)^{m_1+m_2}(R^\dagger)^{N-(m_1+m_2)}|\text{vac}\rangle \\
&= \frac{1}{\sqrt{2^N s!(N-s)!}}\sum_{m_1=0}^{s}\sum_{m_2=0}^{N-s}(-1)^{m_1}\binom{s}{m_1}\binom{N-s}{m_2} \\
&\quad\times \sqrt{(m_1+m_2)!(N-(m_1+m_2))!}|m_1+m_2\rangle. \qquad (11.12)
\end{aligned}
$$

Consequently, $\overline{\langle s|}P_m\overline{|s\rangle}$ is

$$
\overline{\langle s|}P_m\overline{|s\rangle} = \frac{m!(N-m)!}{2^N s!(N-s)!}\left(\sum_{m_1=\max\{m-N+s,0\}}^{\min\{m,s\}}(-1)^{m_1}\binom{s}{m_1}\binom{N-s}{m-m_1}\right)^2.
$$

This sum is tractable according to Mathematica, if one counts hypergeometric functions as tractable. Note that

$$
\overline{\langle s|}P_m\overline{|s\rangle} = |f_{ms}|^2 \qquad (11.13)
$$

with $f_{ms} = \langle m\overline{|s\rangle}$. Then we find

$$
\begin{aligned}
f_{ms} = \sqrt{\frac{m!(N-m)!}{2^N s!(N-s)!}}&\Bigg[\binom{N-s}{m}{}_2F_1(-m,-s,Nt-s-m,-1) \\
&\times\theta[N-(m+s)-1/2] \\
&+(-1)^{N+s+m}\binom{s}{m-N+s}{}_2F_1(m-N,-N+s,1+m-N+s,-1) \\
&\times\theta(-(N-(m+s)-1/2))\Bigg] \qquad (11.14)
\end{aligned}
$$

where $\theta(x)$ is the unit step function. A check is in order: for $s=0$, the expectation becomes $\overline{\langle 0|}P_m\overline{|0\rangle} = \binom{N}{m}/2^N$, as expected for the zero temperature ground state.

Now the thermal expectation is

$$
\langle P_m\rangle = \frac{1}{\sum_s e^{-\beta E_s}}\left(\sum_s e^{-\beta E_s}|f_{ms}|^2\right) \qquad (11.15)
$$

with f_{ms} given by Eq. 11.14. These are plotted in Fig. 11.1.

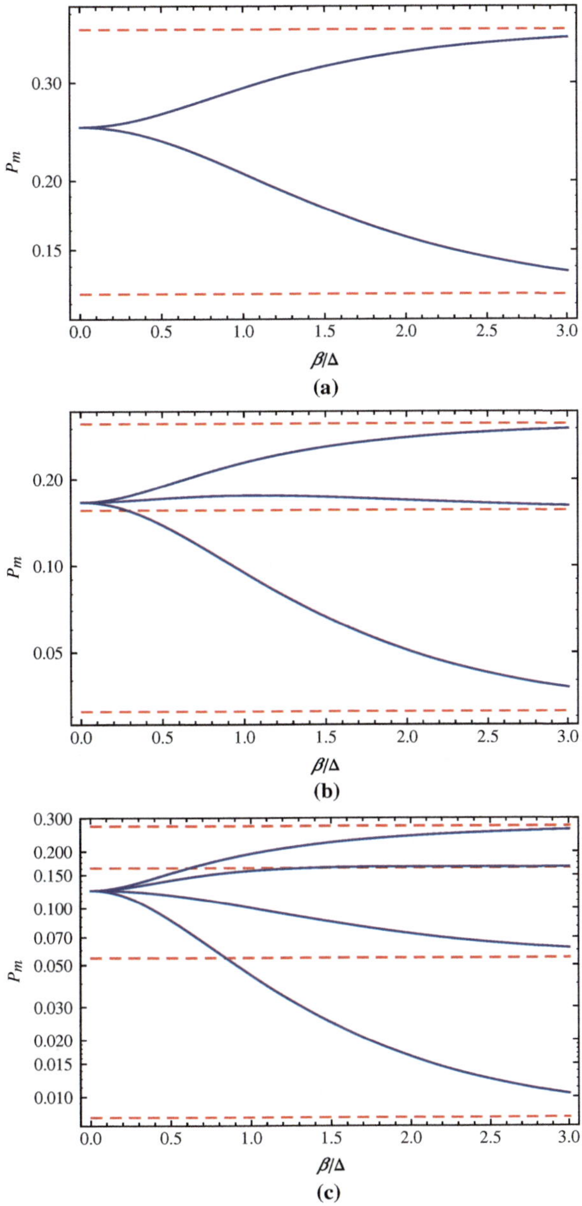

Fig. 11.1 Thermal expectation values $\overline{\langle s|}P_m\overline{|s\rangle}$ versus inverse temperature $\beta\Delta$. At high temperature, all states are equally likely, while at low temperature they tend to a binomial distribution, indicated by the dashed *red lines*. *Top* to *bottom*: $N = 3, 5, 7$. It should be clear from the graphs which curves correspond to which m

11.4.1.3 Operator Correlations of $P_m = |m\rangle\langle m|$

Here we would like to calculate the imaginary time Green's function $\mathcal{G}_{mp}(\tau) = \langle \mathcal{T} P_m(\tau) P_p(0)\rangle$ defined by Eq. 11.5. We will use spectral representation of the imaginary time Green's function, which allows us to obtain $\mathcal{G}_{mp}(\omega_n)$ given the eigenstates and eigenvalues. Note that in this section, rather than denoting many-body states with an overline—as $\overline{|s\rangle}$—I will denote them implicitly by using Greek letters—e.g, $|\alpha\rangle$.

The spectral or Lehmann representation for the thermal imaginary time Green's functions is

$$\mathcal{G}_{mp}(\omega_n) = \frac{1}{\mathcal{Z}} \sum_{\alpha\beta} \left\{ \langle\alpha|P_m|\beta\rangle\langle\beta|P_p|\alpha\rangle \frac{e^{-\beta E_\alpha} - e^{-\beta E_\beta}}{i\omega_N - (E_\beta - E_\alpha)} \right\} \tag{11.16}$$

defining

$$\mathcal{Z} \equiv \sum_\alpha e^{-\beta E_\alpha}. \tag{11.17}$$

The derivation is standard from many-body theory. Using our results in the previous section for $\langle\beta|P_p|\alpha\rangle$ and E_α, this is

$$\mathcal{G}_{mp}(\omega_n) = \frac{1}{\mathcal{Z}} \sum_{\alpha\beta} \left\{ f_{\alpha m} f_{\beta m} f_{\alpha p} f_{\beta p} \frac{e^{-\beta E_\alpha} - e^{-\beta E_\beta}}{i\omega_N - (E_\beta - E_\alpha)} \right\} \tag{11.18}$$

where the $f_{\alpha m}$'s are given by Eq. 11.4 and the E_α's are given by Eq. 11.10. We may further drop the constant in this equation to obtain $E_s = s\Delta$.

11.4.2 Strong Coupling $t/U \to 0$ limit

In this section, I calculate the correlation function $\mathcal{G}_{mn}(\omega_n)$ in the limit of small tunnelings. This will turn out to be quite useful because the expressions obtained are analytically simple and fairly intuitive because the correlations are in a basis of almost-eigenstates.

I have carried out the calculation with two separate techniques, each with their own virtues. In the first method, I have utilized a "slave-boson" technique (this is described below, but is the same basic idea as "Schwinger bosons" in magnetic systems, mentioned just in case one is more familiar with that). This technique enables one to easily pick out and calculate the terms of interest out of the many terms present (e.g., the off-diagonal, non-zero modes of the correlation function). Moreover, the formalism enables the use of familiar many-body perturbation theory, and one can easily go to more sites and higher orders in "t/U"—a very challenging (or at least tedious) task with the more traditional second method. However, there are a few difficulties with the technique: (1) the theory obtains Bose functions where Boltzmann functions should be obtained, (2) there is a free parameter in the theory—the slave boson chemical potential—which seems difficult or impossible to solve for analytically, and (3) some results in the literature seem to indicate that the slave boson approach is inexact at non-zero temperature, even in the strong-coupling limit.

Here, to complete the slave boson calculation despite these difficulties, I overcome its failures by replacing, by hand, the Bose functions with Boltzmann functions, in which case the constant factor may be obtained analytically. Doing this yields the exact solution obtained via our second technique, however given the somewhat *ad hoc* nature of the last steps in this calculation, it is unclear whether the method will work to higher order and for more sites.

In the second method, which suffers none of the stumbling blocks of the slave boson theory, I work directly with the spectral representation for the response function, Eq. 11.16, solving for the many-body eigenstates and eigenvalues perturbatively to obtain the response function to second order in "t/U". Although this has the virtue of having no handwaving steps in the calculation, it is quite tedious already at second order with two sites, and the > 2 site, more than second order calculation would be painful.

Before giving the calculations, I will set up the general formalism. I compute the response to first order in the hoppings by splitting Eq. 10.4 into

$$H = H_0 + \lambda H_{\text{pert}} \tag{11.19}$$

with

$$H_0 = \sum_{i,n} \epsilon_n |n\rangle_i \langle n|_i. \tag{11.20}$$

and

$$H_{\text{pert}} = -\sum_{\langle i,j\rangle;m,n} t_{ij}^{(mn)} |m+1\rangle_i |n-1\rangle_j \langle m|_i \langle n|_j \tag{11.21}$$

and computing properties to first order in the control parameter λ (which is unity, physically). This should be accurate when the $t^{(mn)}$'s are smaller than the energy *differences* between the ϵ_m—that is, for deep lattices.

11.4.2.1 Small Hopping, Calculation with Slave Bosons

The first approach uses slave bosons. The idea is to first note that the Hilbert space on each site j is spanned by the kets $|n\rangle_j$ enumerated by site fillings $n = 0, 1, 2, \ldots$. Then we may introduce a set of creation operators $\phi_n^\dagger$ which act on the vacuum to give these. In an equation, we have:

$$\phi_{j,n}^\dagger |\text{vac}\rangle = |n\rangle_j \tag{11.22}$$

where the $\phi_{j,n}$'s ($\phi_{j,n}^\dagger$'s) are usual bosonic annihilation (creation) operators. Note that the vacuum is *not* the $n = 0$ state. We may then construct our theory in terms of the ϕ operators in the usual manner, obtaining a relatively simple bosonic theory in the Hilbert space formed by allowing spanned by the $(\phi_n^\dagger)^\alpha$'s acting on the vacuum. However, this theory has one unusual aspect: in this prescription we have expanded our Hilbert space enormously—for each ket $|n\rangle$ in the original rpoblem there are now many kets $\phi_n^\dagger|\mathrm{vac}\rangle, (\phi_n^\dagger)^2|\mathrm{vac}\rangle, (\phi_n^\dagger)^3|\mathrm{vac}\rangle, \ldots$ Consequently, in order to reproduce the original Hilbert space we need to enforce the constraint for the particles to remain in the original Hilbert space. This is most easily accomplished in a coherent state path integral formalism giving a generating function $\mathcal{Z}$ for imaginary time correlation functions as

$$\mathcal{Z} = \int \left(\prod_{n=0}^{\infty} \mathcal{D}\phi_n\right) \left[\prod_{\tau,j} \delta\left(\sum_{n=0}^{\infty} \phi_{j,n}^*(\tau)\phi_{j,n}(\tau) - 1\right)\right] \times \exp[-S_{\mathrm{sb}}[\phi^*, \phi]], \tag{11.23}$$

the integral runs over all functions periodic with period β, and the delta function constraint limits our particles to the original Hilbert space—that is, it ensures the system has probability of unity for occupying some state in the on-site Hilbert space; where the action is

$$S_{\mathrm{sb}} = \int_0^\beta d\tau \left\{ \sum_j \phi_{j,n}^*(\tau)\partial_\tau \phi_{j,n}(\tau) + H_{\mathrm{sb}}[\phi_{j,n}] \right\}, \tag{11.24}$$

with slave-boson Hamiltonian

$$H_{\mathrm{sb}}[\phi_{j,n}] = \sum_{i,m} E_{i,m}\phi_{i,m}^*\phi_{i,m} - \sum_{\langle i,j\rangle,mn} t^{(m,n)} \phi_{i,m+1}^* \phi_{j,n-1}^* \phi_{i,m}\phi_{j,n}. \tag{11.25}$$

Equation 11.25 reveals a major strength of the slave boson formalism: it treats the on-site term at quadratic order so the quadratic theory is exact in the $t/U = 0$ limit, while the hopping term now takes the form of a (somewhat unfamiliar) two-body interaction, which may be treated using usual diagrammatic perturbation theory. The sole caveat is that one must enforce the constraint at the end of the calculation; this caveat may be removed by an approximation: we (exactly) rewrite the delta function as

$$\left[\prod_{\tau,j} \delta\left(\sum_n \phi_{j,n}^*(\tau)\phi_{j,n}(\tau) - 1\right)\right] = \int \mathcal{D}\xi \, e^{i\int d\tau \sum_j \xi_j(\tau)\left(\sum_n \phi_{j,n}^*(\tau)\phi_{j,n}(\tau) - 1\right)} \tag{11.26}$$

(dropping constant factors from the integration measure as usual) and follow this by approximating the $\xi_j(\tau)$ as a constant, only enforcing the constraint on average over the system and over time:

$$\int \mathcal{D}\xi\, e^{i\int d\tau \sum_j \xi_j(\tau)\left(\sum_n \phi^*_{j,n}(\tau)\phi_{j,n}(\tau)-1\right)}$$
$$\doteq \int d\xi\, e^{i\xi \int d\tau \sum_j \left(\sum_n \phi^*_{j,n}(\tau)\phi_{j,n}(\tau)-1\right)} \tag{11.27}$$

so that ξ appears only as an effective chemical potential for the slave bosons. I believe this is exact at zero temperature and strong couplings.

At this point, one may do various sophisticated things; I do the simplest, a straightforward perturbation theory in the hopping terms. Because our system is now described by an ordinary bosonic theory once we approximate the constraint with an effective chemical potential, we may use ordinary diagrammatic perturbation theory to calculate the finite temperature correlation functions. We are interested in $\mathcal{G}_{mp}(\nu_n) = \langle TP_m(\tau)P_p(0)\rangle$. For future use, I would like to generalize this to multi site systems. Recall that we defined P_m for the two site, N total particle system as $P_m \equiv |m\rangle_L |N-m\rangle_R \langle m|_L \langle N-m|_R$. For multiple sites, probably the most useful generalization of $\mathcal{G}$ to measure with PIMC is

$$\mathcal{Q}_{mn;ij}(\tau) \equiv \langle TS_{mi}(\tau)S_{nj}(0)\rangle \tag{11.28}$$

with

$$S_{mi} \equiv |m\rangle_i \langle m|_i. \tag{11.29}$$

Written in terms of the slave boson operators, the generalized $\mathcal{Q}$ is

$$\mathcal{Q}_{mp;ij}(\tau) = \langle T\phi^*_{mi}(\tau)\phi_{mi}(\tau)\phi^*_{nj}(0)\phi_{nj}(0)\rangle. \tag{11.30}$$

Some generic diagrams for this are shown in Fig. 11.2: the hoppings now appear as a more or less ordinary two particle interaction. For example, in words diagram (c) has a particle hopping from site m to a site s and then to site j. One strength of this approach is immediate: since the "bare" propagator is diagonal in the on-site Hilbert space, we immediately see that only diagram "(f)" contributes to the off-diagonal response to second order in "t/U."

In general we can evaluate the perturbation series with these diagrams. However, for two sites the problem is even simpler. We return to $\mathcal{G}$ rather than the $\mathcal{Q}$. In this case, rather than introducing slave bosons for each occupation at each site, we take advantage of the fact that once the occupation of the left site is known, the right site's occupation is also known. So we introduce slave boson operators ϕ_m and ϕ^*_m for each state $|m\rangle$, for $m = 0, 1, \ldots, N$. Then the functional integral is similar to before, but the slave boson Hamiltonian is even simpler:

$$H_{sb} = \epsilon_m \phi^*_m \phi_m - t^{(m-1)}\phi^*_m \phi_{m-1} - t^{(m)}\phi^*_m \phi_{m+1} \tag{11.31}$$

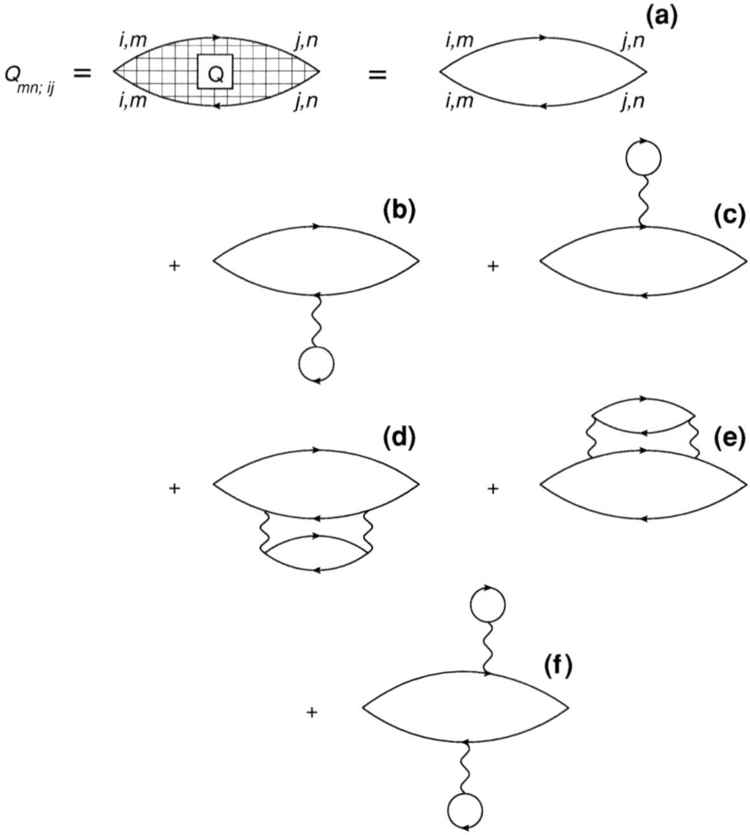

Fig. 11.2 Representative diagrams for the slave boson perturbation theory at low order in "t/U"

with $\epsilon \equiv E_m + E_{N-m}$. The diagrammatic theory for this problem is then simply that of a particle in an external potential. From this, we can easily calculate $\mathcal{G}_{mn}(\tau) = \langle T\phi_m(\tau)\phi_m(\tau)\phi_n(0)\phi_n(0)\rangle$, writing the Green's function in terms of the slave boson operators, as illustrated in Fig. 11.3.

Out of all the figures in Fig. 11.3 only (f) contributes to the off-diagonal response. Using standard diagrammatic rules, it gives the second order correction to the off-diagonal piece $\delta\mathcal{G}^{(2)}_{mp}$ to be

$$\delta\mathcal{G}^{(2)}_{mp}(v_n) = \frac{1}{\beta^2}\sum_{\omega_1,\omega_3}\frac{1}{(-i(v_n+\omega_1)-\epsilon_m)(-i(v_n+\omega_3)-\epsilon_n)} \times \frac{\delta_{\omega_1,\omega_3}\left(-t^{(m-1)}\delta_{m-1,n}-t^{(m)}\delta_{m+1,n}\right)^2}{(-i\omega_1-\epsilon_m)(-i\omega_3-\epsilon_n)}. \tag{11.32}$$

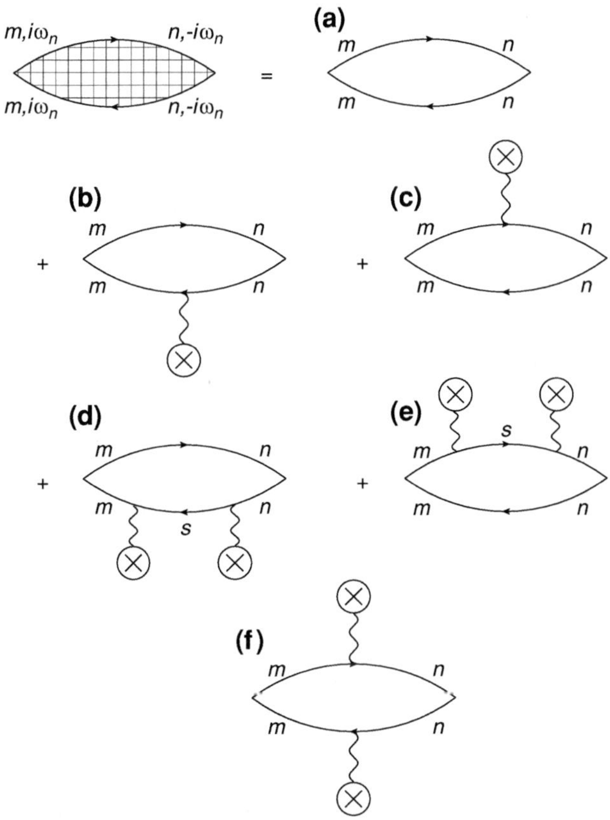

Fig. 11.3 Representative diagrams for the two-site slave boson perturbation theory at low order in "t/U"

One may evaluate the Matsubara sums in the usual fashion, giving

$$\delta\mathcal{G}^{(2)}_{mn}(\nu_n) = \frac{2\left(t^{(m)}\right)^2 \delta_{m+1,n}}{(\epsilon_m - \epsilon_{m+1})\beta} \frac{n(\epsilon_{m+1}) - n(\epsilon_m)}{\nu_n^2 + (\epsilon_{m+1} - \epsilon_m)^2} + \frac{2\left(t^{(m-1)}\right)^2 \delta_{m-1,n}}{(\epsilon_m - \epsilon_{m-1})\beta} \frac{n(\epsilon_{m-1}) - n(\epsilon_m)}{\nu_n^2 + (\epsilon_{m-1} - \epsilon_m)^2} \tag{11.33}$$

where $n(\epsilon) \equiv \left(e^{\beta(\epsilon-\xi)} - 1\right)^{-1}$ is the usual Bose function (no longer absorbing the chemical potential into the energies ϵ).

We have yet to determine the slave bosons' chemical potential ξ. This requires solving

$$\sum_m n(\epsilon_m) = 1 \tag{11.34}$$

which it wasn't obvious to me how to do, even for $t^{(n)} = 0$. However, an *ad hoc* guess reproduced the exact answer (calculated below): we replace the Bose functions with Boltzmann functions, in which case the chemical potential is simply a normalization, which is $1/\sum_m e^{-\beta\epsilon_m}$. This exact answer compares favorably with the exact diagonalization and is confirmed doing direct perturbation theory below.

There is one point that must be kept in working with Eq. 11.33 in practice: when energies m and $m+1$ or m and $m-1$ approach each other, one must take the limit of the summand where the energies approach each other to avoid the indeterminate $0/0$.

11.4.2.2 Small Hopping Calculation with Ordinary Perturbation Theory in Spectral Representation

In this section I calculate the correlation function $\mathcal{G}_{mn}(\omega_n)$ in the limit of small tunnelings directly from the spectral representation, which is more tedious and less intuitive than the slave bosons, but doesn't require any "handwavey" steps.

We start with the spectral representation given in Eq. 11.16 and expand the eigenstates and eigenvalues to second order in t/U. Careful examination reveals that one only needs to expand the eigenstates to first order in t/U and can neglect the expansion of the eigenenergies. In a little detail: (1) the energy corrections occur to lowest order at t^2, so for second order terms, the many-body states in the matrix elements $\langle\alpha|P_m|\beta\rangle\langle\beta|P_p|\alpha\rangle$ may be taken to be the unperturbed states. In this case, the response function gets only a diagonal contribution, which is uninteresting for us. (2) The second order corrections to the many-body eigenstates also give only diagonal contributions. So we simply calculate the matrix elements in the spectral representation by computing the many-body eigenstates' corrections to lowest order. Naively, this gives 24 terms which I don't write down here. Of these, only four terms survive, and the spectral spectral representation gives

$$\delta\mathcal{G}^{(2)}_{mp}(\nu_n) = \frac{1}{\sum_m e^{-\beta\epsilon_m}}\left[\frac{\left|t^{(m)}\right|^2\delta_{m,n-1}}{(\epsilon_m-\epsilon_{m+1})^2}\left(\frac{e^{-\beta\epsilon_{m+1}}-e^{-\beta\epsilon_m}}{i\nu_n-(\epsilon_m-\epsilon_{m+1})}+\frac{e^{-\beta\epsilon_m}-e^{-\beta\epsilon_{m+1}}}{i\nu_n-(\epsilon_{m+1}-\epsilon_m)}\right)\right.$$
$$\left.+\frac{\left|t^{(m-1)}\right|^2\delta_{m,n+1}}{(\epsilon_m-\epsilon_{m-1})^2}\left(\frac{e^{-\beta\epsilon_{m-1}}-e^{-\beta\epsilon_m}}{i\nu_n-(\epsilon_m-\epsilon_{m-1})}+\frac{e^{-\beta\epsilon_m}-e^{-\beta\epsilon_{m-1}}}{i\nu_n-(\epsilon_{m-1}-\epsilon_m)}\right)\right] \quad (11.35)$$

which simplifies to

$$\delta\mathcal{G}^{(2)}_{mp}(\nu_n) = \frac{2}{\sum_m e^{-\beta\epsilon_m}}\left[\delta_{m,n-1}\left(e^{-\beta\epsilon_{m+1}}-e^{-\beta\epsilon_m}\right)\frac{\left(t^{(m)}\right)^2}{(\epsilon_{m+1}-\epsilon_m)\left(\nu_n^2+(\epsilon_m-\epsilon_{m+1})^2\right)}\right.$$
$$\left.+\,\delta_{m,n+1}\left(e^{-\beta\epsilon_{m-1}}-e^{-\beta\epsilon_m}\right)\frac{\left(t^{(m-1)}\right)^2}{(\epsilon_{m-1}-\epsilon_m)\left(\nu_n^2+(\epsilon_m-\epsilon_{m-1})^2\right)}.\right] \quad (11.36)$$

This agrees very well with the exact result and agrees exactly with the slave boson calculation supplemented with the Bose→Boltzmann replacement.

11.4.3 General—Correlation Functions from Exact Diagonalization

We want to find the correlation functions $\mathcal{G}_{mp}(\omega_n)$ for our 2-site generalized Bose-Hubbard model, described by the Hamiltonian of Eq. 10.4. Rewriting this for two sites, using our now-familiar notation $|m\rangle \equiv |m\rangle_L |N-m\rangle_R$, we obtain

$$\langle m|H|n\rangle = -t^{(m-1,N-m+1)}\delta_{m-1,n} - t^{(N-m-1,m+1)}\delta_{m+1,n} + (\epsilon_m + \epsilon_{N-m})\delta_{mn} \quad (11.37)$$

We may now find the $\mathcal{G}_{mn}(\omega_n)$'s by finding this matrix's eigenvalues and eigenvectors, and then using $\mathcal{G}_{mn}$'s spectral representation Eq. 11.16.

Note that because of the symmetry under interchanging α and β, the expressions for $\mathcal{G}$ may look nice by rewriting the $(i\omega - E_\alpha + E_\beta)^{-1}$ in terms of a Lorentzian in both this and previous sections.

11.5 Note on Temperature Dependence of Response Functions

At high temperature, temperature gives only an overall scale: the exponential difference in spectral expansion expands so that one gets only a constant factor of $1/T$ (and the scale of my spectral functions does change, consistent with this; there is also a change in scale from the partition function). At low temperature, the temperature dependence again gives only an overall scale: the ground state should remain in the sum, others disappear, by which I mean one sum disappears (either α =GS and β sums or β =GS and α sums)—so $e^{-\beta E_{GS}}$ factors. Similarly, non-interacting Green's functions will be temperature independent except for a possible scale. So it is unsurprising if the shape of our imaginary time correlation functions are roughly independent of temperature. This is a completely general property of Green's functions, not just the ones we have considered here.

Reference

1. D.M. Ceperley, Rev. Mod. Phys **67**, 279 (1995)

Chapter 12
Summary, Conclusions, and the Future of Induced Gauge Fields and Lattices with On-Site Correlations

Although rotation and effective magnetic fields have been around essentially since the first observations of Bose–Einstein condensates, they have only more recently become capable of approaching exotic states. The (former) Chu group have created fractional quantum Hall puddles in a lattice, using methods similar to those discussed in Chap. 9 (N. Gemelke, private communication). The NIST group have recently demonstrated creation of vortices via light-induced gauge fields, and these are estimated to be capable of reaching the $\sim$100 particle quantum Hall regime in elongated geometries [1].

Theoretically, it would be interesting to better understand the connection between these arrays of quantum Hall puddles in lattices where tunneling is important (as discussed in Chap. 10) and networks of FQH states in semiconductors [2]. Experimentally, this physics can be explored in the immediate future by continuing the Chu group's experiments mentioned above at lower lattice depths.

The importance of on-site correlations in other situations has been experimentally demonstrated, even in seemingly weakly interacting ^{87}Rb optical lattice experiments. It will be interesting to quantitatively examine these and other experiments' inferred model parameters with those calculated using the techniques of Chap. 11. There are further interesting opportunities for using lattices far outside the Hubbard regime to maximize the transition temperature of desired ordered states (such as antiferromagnetism) [3] or to explore universal phenomena that are insensitive to these microscopic details (such as quantum criticality, see Chap. 13).

References

1. Y. Lin, R.L. Compton, K. Jimenez-Garcia, J.V. Porto, I.B. Spielman, Nature **462**, 628 (2009)
2. E. Shimshoni, A. Auerbach, A. Kapitulnik, Phys. Rev. Lett. **80**, 3352 (1998)
3. L. Mathey, E. Altman, A. Vishwanath, Phys. Rev. Lett. **100**, 240401 (2008)

K. R. A. Hazzard, *Quantum Phase Transitions in Cold Atoms and Low Temperature Solids*, Springer Theses, DOI: 10.1007/978-1-4419-8179-0_12,

Chapter 13
Techniques to Measure Quantum Criticality in Cold Atoms

13.1 Introduction

Attempts to understand zero temperature phase transitions have forced physicists to consider a regime where the standard paradigms of condensed matter physics break down [1–4]. These quantum critical systems lack a simple description in terms of weakly interacting quasiparticles, but over the past 20 years physicists have gained deep insights into their properties. Most dramatically, theory predicts that universal scaling relationships describe their finite temperature thermodynamics up to remarkably high temperatures. Unfortunately, these universal functions are hard to calculate: for example there are no reliable general techniques [4, 5] to calculate the scaling functions for dynamics. Viewing a cold atom experiment as a quantum simulator [6], we show how to extract universal scaling functions from (non-universal) atomic density profiles. Such experiments can resolve the important open question of the Mott-Metal crossover's dynamic exponent in the Fermi-Hubbard model [7, 8] and explore the poorly characterized crossover between $O(2)$ and dilute Bose gas physics captured by the finite density $O(2)$ rotor model [1, 9]. Such results would directly impact theories of, for example, high temperature superconducting cuprates [10, 11], heavy fermion materials [12], and graphene [13].

Although our ideas are general, much of our discussion will focus on cold bosonic atoms, such as ^{87}Rb, trapped in optical lattices. Current experiments on this system are capable of observing quantum critical phenomena. This system displays multiple quantum phase transitions, which reside in distinct and non-trivial universality classes [1]. We will also discuss a number of other experimental realizations with richer quantum critical physics. Our protocols can be applied to any quantum phase transition.

This chapter was adapted from "Techniques to measure quantum criticality in cold atoms" by Kaden. R. A. Hazzard, Erich. J. Mueller, preprint available at arxiv:1006.0969 (2010).

K. R. A. Hazzard, *Quantum Phase Transitions in Cold Atoms and Low Temperature Solids*, Springer Theses, DOI: 10.1007/978-1-4419-8179-0_13,

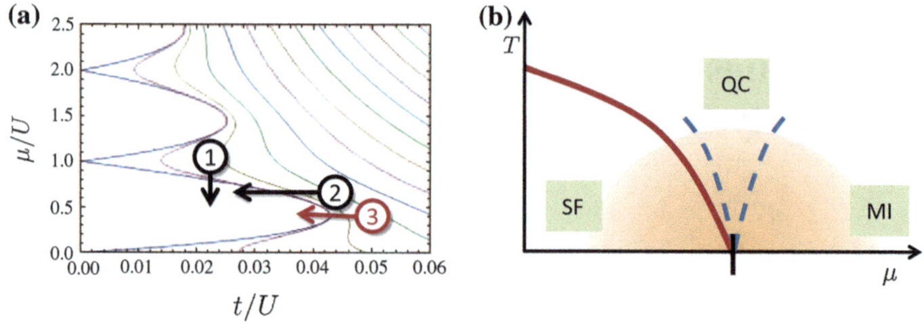

Fig. 13.1 *Quantum critical crossovers in the Bose-Hubbard model.* **a** Mott insulator to superfluid phase boundaries for (*left* to *right*) temperatures $T/U = 0.00, 0.06, 0.12, \ldots, 0.96$, calculated via finite temperature Gutzwiller theory (see Chap. 14) in $d = 2$. The parameters t, μ, and U are the tunneling rate, chemical potential, and on-site interaction energy of the Bose-Hubbard model. Paths (1,2) are governed by the dilute Bose gas universality class, and path (3) (passing through the tip) is governed by the $O(2)$ universality class. See Chap. 14 for more details. **b** Slice through the quantum phase transition at fixed t/U. Slices with fixed μ/U are similar. The abbreviations "SF", "MI", and "QC" denote the superfluid, Mott insulator-like normal fluid, and quantum critical regime, with the Mott insulator being strictly defined only at $T = 0$. The shaded region indicates where the physics is governed by the quantum critical point. The dashed lines represent smooth crossovers between the qualitatively distinct "QC" and "low temperature" regions. Deep in the former, T is the only relevant energy scale

Ultracold bosons in optical lattices are described by the Bose-Hubbard model [6]. Figure 13.1a illustrates this model's zero-temperature phase diagram with a superfluid phase when the tunneling matrix element t is large compared to the on-site interaction U, and a Mott insulating phase in the opposite limit. The superfluid phase is characterized by dissipationless mass transport, analogous to the dissipationless charge transport in a superconductor. It supports arbitrarily low-energy excitations and has an order parameter ψ that vanishes at the phase transition. The Mott insulating phase (lobes shown in Fig. 13.1a) can be caricatured as having a fixed integer number of particles per site, n, set by the chemical potential μ. The Mott insulating state has a excitation gap Δ, which vanishes at the transition.

The Mott insulating phase is strictly defined only at zero temperature, but a phase transition between a superfluid and a normal fluid persists to finite temperature (Fig. 13.1a). Figure 13.1b illustrates a slice through the finite temperature phase diagram. This figure shows that the normal fluid is divided into three qualitatively distinct regions separated by smooth crossovers illustrated by dashed lines. At large t/U, the superfluid (SF) can be heated through a classical phase transition to a normal fluid (NF) with properties similar to that of a weakly interacting Bose gas. At small t/U the finite temperature properties of the normal fluid (MI) are determined by the zero temperature Mott insulator's gap. A quantum critical region intervenes between these two normal fluids. Slices at fixed t/U look identical to slices at fixed μ/U. In fact, this structure is typical of all second-order quantum phase transitions: the disordered phase generically divides into these three regions [1].

Sufficiently near the quantum phase transition (the shaded region in Fig. 13.1b), the thermodynamic functions obey scaling relationships. This scaling is typically controlled by two energy scales that vanish at the quantum critical point [1]. One scale is the temperature T, while the other depends on the phase: it is the zero temperature superfluid stiffness ρ_s in the superfluid and the gap to single particle excitations Δ in the Mott phase. Observables, such as the density, can be written as the sum of an non-universal part n_0 plus a function n_u that obeys the scaling form

$$n_u(T, \Delta) = T^{d/z}\Psi_n(\Delta/T). \tag{13.1}$$

Here d is the dimension of the system and Ψ_n is a universal scaling function. Below an upper critical dimension d_c, the dynamical exponent z and the universal function Ψ_n are the same for a wide range of models and therefore define universality classes: here $d_c = 4 - z$, and z will be 1 or 2 depending on what part of the Mott lobe one is traversing. The non-universal contribution, $n_0(\mu, U, T)$, is the density of the system for $t = 0$, and as explained in Chap. 14 is easily calculated or measured in independent experiments. Although atypical, with fine tuning it is possible to have situations where additional dimensionless parameters enter. The finite density $O(2)$ rotor model discussed below is one example.

Here we introduce an experimental protocol to observe this universality and measure key properties such as the dynamic critical exponent. Our technique relies upon the inhomogeneity of cold atoms experiments. Although typical observables in cold atoms, such as the density profile, are non-universal, we show that with appropriate analysis the absorption images of these trapped inhomogeneous clouds reveal the universal properties of corresponding homogeneous systems. We find an additional remarkable result: universal quantum critical scaling persists to much smaller clouds than the features associated with the finite temperature *classical* transitions (Fig. 13.1a, solid line), which have been theoretically and experimentally studied [14–17].

We will view trapped atomic systems as locally satisfying Eq. 13.1, with a gap Δ that varies as a function of position in space. As discussed below, our analytic and numerical results in trapped systems validate this Thomas-Fermi approximation. Direct measurement of $\Delta(\mathbf{r})$ is difficult in cold atoms. Instead, we will relate the gap to independently measurable thermodynamic quantities.

We illustrate our procedure by taking the compressibility $\kappa \equiv \partial n/\partial\mu$ as the independently measurable quantity. This is appealing because κ can be extracted directly from the density profiles via $\kappa = -(1/m\omega^2 r)\partial n/\partial r$ or from n's shot-to-shot fluctuations [18]. Like the density, the compressibility is the sum of an easily determined non-universal part and a universal part, $\kappa = \kappa_0 + \kappa_u$, with

$$\kappa_u(T, \Delta) = T^{d/z-1}\Psi_\kappa(\Delta/T). \tag{13.2}$$

Equation 13.1 is an implicit equation for Δ which together with Eq. 13.2 implies that

$$\kappa_u T^{1-d/z} = \Upsilon(n_u T^{-d/z}) \tag{13.3}$$

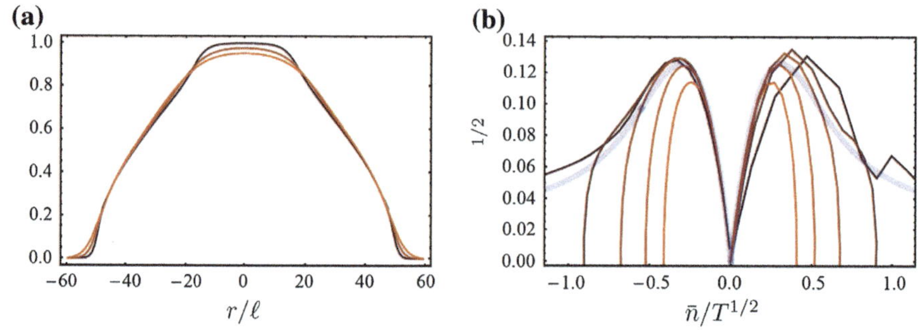

Fig. 13.2 *Extracting universal behavior and dynamical critical exponents from density profiles of the trapped one-dimensional Bose-Hubbard model.* **a** Exact density profiles of the one dimensional harmonically trapped hard core Bose-Hubbard model for $N = 70$ particles at temperatures $\bar{T} \equiv T/t = 0.1, 0.24, 0.38$, with larger temperatures corresponding to lower central density. Here, r is the radial displacement in the trap and d is the lattice spacing. These density profiles are non-universal: for example, they depend on temperature. **b** Our construction for obtaining universal scaling curves applied to this system, plotting $\bar{\kappa}\bar{T}^{d/z}$ versus $\bar{n}/\bar{T}^{1/2}$ (defining $\bar{\kappa} \equiv \kappa - \kappa_0$, $\bar{n} \equiv n - n_0$, and $\bar{T} = T/t$) for this $d = 1$, $z = 2$ transition and temperatures $\bar{T} = 0.1, 0.17, 0.24, 0.31, 0.38$. The compressibility is approximated by $\kappa = \partial n/\partial\mu \approx (\partial n/\partial r)/(m\omega^2 r)$, where ω is the trap frequency, and $\partial n/\partial r$ is obtained by numerically differentiating the density. Lower temperatures display a larger region of collapse. We observe good collapse up to $T \sim 0.25t$, and that the analysis accurately reproduces the homogeneous infinite system's scaling curve (*shaded gray line*) within $\lesssim$10% for $T \gtrsim 0.05t$ for the transition near $n = 1$ ($\bar{n} < 0$) and for $T \gtrsim 0.15t$ for the transition near $n = 0$ ($\bar{n} > 0$). This collapse occurs even for drastically different density profiles obtained by adjusting the trap depth in place of temperature (not shown). With moderately larger particle numbers ($N \sim 200$, not shown), the simulated data at low temperatures even more accurately reproduces the infinite homogeneous system's universal scaling function (the extracted curve lies within the *shaded gray* region)

where $\Upsilon = \Psi_\kappa \circ \Psi_n^{-1}$ is a universal function. Therefore, in the quantum critical regime a graph of $\kappa_u T^{1-d/z}$ versus $n_u T^{-d/z}$ collapses onto a single curve independent of μ, t, U, T, and any other microscopic couplings.

Figure 13.2b shows this plot extracted from the density profiles (Fig. 13.2a) of the $t/U \to 0$ trapped one-dimensional Bose-Hubbard model for $N = 70$ particles. These density profiles are efficiently calculated by mapping the system onto non-interacting fermions (see Chap. 14) The figure not only demonstrates the universal collapse in the quantum critical regime, but also validates the Thomas-Fermi approximation. The success of the Thomas-Fermi approximation is due to the fact that as one increases temperature above the quantum critical point the coherence length ξ shrinks, and throughout the quantum critical regime it is typically much smaller than the cloud size. As one approaches the classical phase transition, ξ diverges. Hence even though our technique quite readily detects quantum criticality, signatures of the classical transition may be much harder to see [17].

As shown in Fig. 13.2b, collapse occurs in a region around the quantum critical point, but deviates at sufficiently large densities and temperatures. In this case, the data collapses within 10% for a density range of roughly $0.9 < n < 1.1$ and

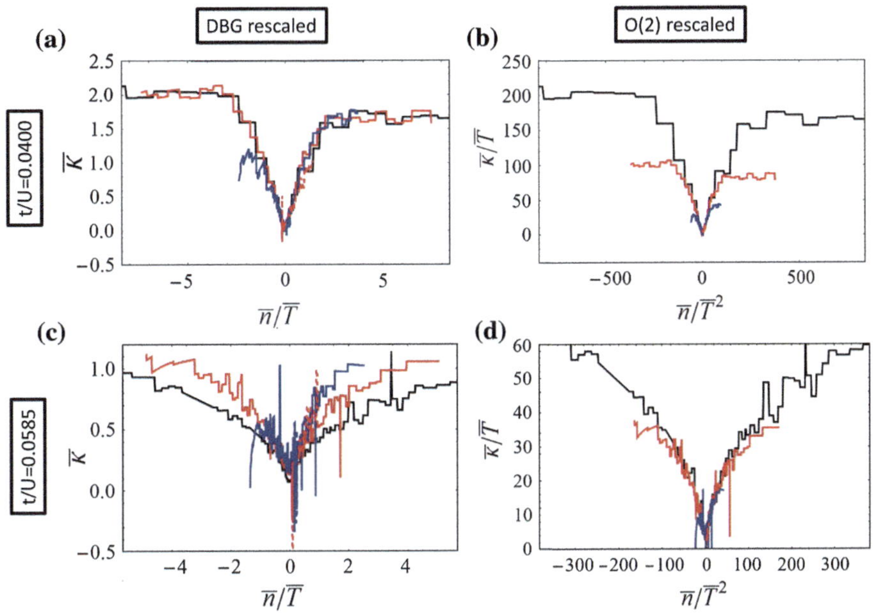

Fig. 13.3 *Extracting universal behavior and dynamical critical exponents from density profiles of the trapped two-dimensional Bose-Hubbard model.* Each panel shows an application of our analysis procedure to simulated density profiles (as described in text) for temperature $T/t = 1/4, 1/2, 1, 2,$ and 4 using the data in a density range $|n-1| < 0.15$. The symbols $\bar{n}$, $\bar{\kappa}$, and $\bar{T}$ are defined in Fig. 13.2. Lower temperature curves are identified by noting that they span a wider $\bar{n}$ range. On the *left* (**a** and **c**), we plot $\bar{\kappa}$ versus $\bar{n}\bar{T}^{-1}$, while on the *right* (**b** and **d**) we plot $\bar{\kappa}\bar{T}^{-1}$ versus $\bar{n}\bar{T}^{-2}$: these will show collapse if the dynamical critical exponent is respectively $z = 2,1$. *Top* (**a** and **b**) shows data with $t/U = 0.0400$, which should be described by the dilute Bose gas universality class ($z = 2$). *Bottom* (**c** and **d**) shows data with $t/U = 0.0585$, which should be better described by the $O(2)$ rotor model universality class ($z = 1$): the tip of the $n = 1$ Mott lobe in the homogeneous system is at $t/U = 0.0593$. As expected, we see collapse in **a** and **d**. The scatter in data points corresponds to stochastic noise in our Monte-Carlo simulations amplified by the differentiation. This noise is of comparable size to what would be seen in an experiment

temperatures range of $T \lesssim t/4$. This is a generic feature of quantum criticality: the physics is universal only near the quantum critical point, with the size of the universal region determined by microscopic details.

The universal scaling functions in Fig. 13.2 correspond to those of the dilute Bose gas universality class (the same universality class which governs the edge of all cold bosonic clouds). Although this universality class is quantitatively understood [19, 1], at larger t/U (near the tip of the lobe) the Bose-Hubbard model falls into the less trivial $O(2)$ universality class, for which open questions remain. In particular, the quantitative structure of the dynamic scaling functions is unknown and no general methods exist to compute these. Furthermore, there is a more intricate "finite density $O(2)$" universal structure governing the crossover between the $O(2)$ and dilute Bose gas universality classes, about which even less is known [9, 13]. We give a simple picture of this physics and its relevance to the Bose-Hubbard model in Chap. 14

Figure 13.3 shows that in addition to measuring universal scaling functions and the size of the universal region, our method enables one to distinguish the universality classes of the two-dimensional square lattice Bose-Hubbard model using only the density profiles. We calculate the equation of state $n(\mu)$ for a set of μ's realized in a typical cold atoms experiment, employing the numerically exact worm algorithm quantum Monte Carlo as implemented in the ALPS package [20]. To mimic the effect of the trapping potential, we calculated each density for a finite size 10×10 system. This length scale corresponds to a trapping potential variation of $\sim 1\%$ along the radial direction under typical conditions. We calculate $\kappa(\mu)$ using the lowest order finite difference derivative on this data. The stochastic error in the quantum Monte Carlo yields results with some noise ($\sim 0.1\%$), similar to that found applying finite differences to real experimental data ($\sim 1\%$ noise in radially averaged density profiles, e.g. in Gemelke et al. experiments [18]; see Chap. 14 for a discussion of signal-to-noise issues).

We are able to distinguish these universality classes because they are characterized by different dynamical critical exponents ($z = 2$ for the dilute Bose gas and $z = 1$ for the $O(2)$ model) [1]. One sees collapse only when the correct critical exponent is used. Like Fig. 13.2, the collapse occurs only sufficiently close to the critical point. The behavior persists up to temperatures on the order of $\gtrsim t$ and for a density range of roughly $0.9 < n < 1.1$. Since we are at the upper critical dimension of the dilute Bose gas model, we expect logarithmic corrections to scaling that are hard to identify on these plots. Consistent with our discussion of the Thomas-Fermi approximation's validity, there is no clear signature of the classical phase transition in these graphs.

Applying this method to Fermi lattice systems [21, 22], for example those emulating the two-dimensional square lattice Fermi-Hubbard model, would allow one to resolve a number of important open questions. The first such question is the nature of the Mott-metal crossover in the Fermi Hubbard model. Many believe that in real materials, such as transition metal oxides, this crossover is a manifestation of a preempted quantum critical regime and should display universal physics analogous to that of the Bose-Hubbard model's normal fluid to Mott insulator crossover (see the discussion in Refs. [1, 7]). If one analyzes the square lattice fermion density profiles with the procedure we have introduced and sees collapse onto universal curves, it would be compelling evidence that in the Fermi-Hubbard model the crossover is governed by a quantum critical point. Although most current experiments [21, 22] are three-dimensional, and thus likely above the upper critical dimension, signatures of scaling (only with mean field exponents) would remain if this quantum criticality is the correct description of the crossover. If the quantum critical regime persists to the same temperature scales as the Bose-Hubbard model, then current atomic experiments are already sufficiently cold to determine if the universal Mott-metal crossover scenario is correct.

Beyond determining if the system is critical, this analysis would also provide the dynamical critical exponent: a quantity whose value is currently unknown for many models, including the 2D square lattice Fermi-Hubbard model. Depending on the character of the metallic state, some filling-controlled Mott-metal transitions display

$z = 2$, while others display $z = 4$ [7]. To determine this exponent one would analyze two-dimensional lattice fermions' density profiles with our scaling techniques: only when z is chosen correctly will collapse occur, and thus our protocol is capable of distinguishing between the various theoretical scenarios. As we already argued, the present generation of experiments is already in the correct regime to conduct such studies.

One can explore an even broader range of open questions if in addition to measuring density profiles one also employs spectroscopic probes to measure dynamical response functions. For example, for lattice systems experimentalists have successfully measured modulation spectra [23], Bragg spectra [24], and radio-frequency spectra [25]. In the quantum critical regime, these response functions satisfy universal scaling forms $\chi(\omega, T, \Delta) = T^{\eta}\Psi_{\chi}(\Delta/T, \omega/T)$. In both fermionic and bosonic systems the details of these dynamic scaling functions are largely unknown, especially the low frequency dynamic behavior and how it connects to higher frequency behavior. Note, however, that the Thomas-Fermi approximation's validity for spectra in the quantum critical regime has not been explicitly checked in our work, and its validity as well as other issues such as sufficient signal-to-noise remain open questions.

Up to this point, all of our discussion has focused on what can be achieved with present experimental capabilities. As experiments reach lower temperatures, our analysis technique can be used to probe ever more fundamental physics. For example, one of the simplest symmetry breaking transitions of a Fermi liquid, that to a spin density wave, is ill-understood: beyond the Hertz-Millis theory, there exist an infinite number of marginally relevant coupling constants and its universality class is unclear [26, 27]. By measuring density profiles in a system displaying similar quantum phase transitions, such as dipolar fermions' transition to a Wigner crystal (charge density wave) or near-resonant fermions transition to an "FFLO" state (charge density wave + superfluid order), our techniques can measure z, and thus provide valuable information about these transitions' universality classes. Additionally, the competing instabilities (e.g. to d-wave superconductivity, nematicity, etc.) seen in transition metal oxides have analogs in lattice fermion experiments, and offer an important and even richer set of open questions regarding quantum critical behavior and "avoided quantum criticality" [2, 3, 5, 4, 28]. Finally, Sachdev and Müller have argued that aspects of the dynamics of the $O(2)$ model studied in the Bose-Hubbard model are related to dynamics near the Dirac point in graphene [13], and cold atoms may probe (among other things) the general relations between universal constants that should appear in both cases.

13.2 Note and Summary

Shortly after completion of our manuscript, Zhou and Ho [29] introduced a closely related technique to observe quantum criticality in cold atoms using density profiles. Given the similarities, we feel compelled to enumerate the differences between our works. This comparison also provides an opportunity to summarize our results.

The main purpose of both our paper, and Zhou and Ho's, is to present techniques for observing universal collapse near a quantum critical point. We advocate plotting density vs. compressibility, while Zhou and Ho recommend density vs chemical potential. We feel that the primary benefit of our method is that both n and κ are local observables. In contrast, μ can only be determined if one knows the central chemical potential μ_0. In principle, μ_0 can be extracted by fitting the density profile at the edge of the cloud where the equation of state is simple. In practice there are technical issues with this approach, including the fact that one must have the *entire cloud* in global equilibrium. In our method one only needs equilibrium in the part of the cloud contributing to the critical region. As a technical aside, Zhou and Ho's scaling prefactors (powers of temperature) look more complicated than ours, however, as discussed in Chap. 14, since density is a conserved quantity their results reduce to ours. For other observables, one must use their more general form.

In order to show that our techniques are practical, both our paper, and Zhou and Ho's, use numerical simulations to model a trapped gas. These studies show that finite size effects are small in current experiments, and signal to noise levels are sufficiently high. While both papers include discussion of the $d = 1$ Bose-Hubbard model in the hard core limit, ours also looks at $d = 2$ and $U < \infty$. This lets us demonstrate that our technique can distinguish different universality classes, and explore the interplay between $O(2)$ and dilute Bose gas physics.

Finally, our paper gives a number of concrete and non-trivial open questions involving quantum criticality which can be studied in the near term by cold gas experiments. These include determining the dynamical critical exponent for the Mott-metal crossover, the dynamics of the finite density $O(2)$ model, and the validity of Hertz-Millis theory to describe various symmetry breakings of strongly interacting Fermi liquids.

Zhou and Ho's paper also contains important results that are not found in our paper—namely a study of finite size scaling, a procedure to find the critical chemical potential of quantum phase transitions, and some discussion of classical phase transitions. We found that reading their paper also helped us view technical aspects of our procedure in a new light and helped us improve our approach to subtract off the non-universal contribution to the density.

Acknowledgements We thank Mukund Vengalattore, Stefan Baur, Stefan Natu, Sourish Basu, Srivatsan Chakram, Eun-Ah Kim, Cheng Chin, Nate Gemelke, Ben Machta, Qi Zhou, Jason Ho, Ed Taylor, Victor Gurarie, Nikolay Prokofiev, Lode Pollet, and Subir Sachdev for discussions. This work was supported by the National Science Foundation through Grant PHY-0758104 and by a grant from the Army Research Office with funding from the DARPA OLE program.

References

1. S. Sachdev, Quantum phase transitions (Cambridge University Press, Cambridge 2001)
2. S.L Sondhi, S.M. Girvin, J.P. Carini, D. Shahar, Rev. Mod. Phys **69**, 315 (1997), copyright (C) 2009 The American Physical Society; Please report any problems to prola@aps.org.

3. M. Vojta, Rep. Prog. Mod. Phys. **66**, 2069 (2003)
4. S. Sachdev, in *Understanding Quantum Phase Transitions*, ed. by L.D. Carr (Taylor & Francis, Boca Raton, 2010), arxiv:0910.1139
5. P. Coleman, A.J. Schofield, Nature **433**, 226 (2005)
6. I. Bloch, J. Dalibard, W. Zwerger, Rev. Mod. Phys. **80**, 885 (2008)
7. M. Imada, A. Fujimori, Y. Tokura, Rev. Mod. Phys. **70**, 1039 (1998)
8. A. Georges, G. Kotliar, W. Krauth, M.J. Rozenberg, Rev. Mod. Phys. **68**, 13 (1996)
9. A.V. Chubukov, S. Sachdev, J. Ye, Phys. Rev. B **49**, 11919 (1994)
10. A. Sokol, D. Pines, Phys. Rev. Lett. **71**, 2813 (1993)
11. S. Sachdev, Rev. Mod. Phys. **75**, 913 (2003)
12. P. Gegenwart, Q. Si, F. Steglich, Nat. Phys. **4**, 186 (2008)
13. S. Sachdev, M. Mueller, J. Phys. Condens. Matter **21**, 164216 (2009)
14. Q. Zhou, Y. Kato, N. Kawashima, N. Trivedi, Phys. Rev. Lett. **103**, 085701 (2009)
15. M. Köhl, T. Donner, S. Ritter, T. Bourdel, A. Öttl, F. Brennecke, T. Esslinger, Advances in Solid State Physics, (Springer-Berlin, Heidelberg, 2008) pp. 79–88.
16. S. Trotzky, L. Pollet, F. Gerbier, U. Schnorrberger, I. Bloch, N.V. Prokofév, B. Svistunov, M. Troyer, Nat. Phys. **6**, 998–1004 (2010). doi:10.1038/nphys1799
17. L. Pollet, N.V. Prokof'ev, B.V. Svistunov, Phys. Rev. Lett. **104**, 245705 (2010)
18. N. Gemelke, X. Zhang, C. Hung, C. Chin, Nature 460:995(2009)
19. S. Sachdev, E.R. Dunkel, Phys. Rev. B **73**, 085116 (2006)
20. A. Albuquerque, F. Alet, P. Corboz, P. Dayal, A. Feiguin, S. Fuchs, L. Gamper, E. Gull, S. Gürtler, A. Honecker, R. Igarashi, M. Körner, A. Kozhevnikov, A. Läuchli, S. Manmana, M. Matsumoto, I. McCulloch, F. Michel, R. Noack, G. Pawlowski, L. Pollet, T. Pruschke, U. Schollwöck, S. Todo, S. Trebst, M. Troyer, P. Werner, S. Wessel, J. Magn. Magn. Mater. **310**, 1187 (2007)
21. R. Joerdens, N. Strohmaier, K. Günter, H. Moritz, T. Esslinger, Nature **455**, 204 (2008)
22. U. Schneider, L. Hackermller, S. Will, T. Best, I. Bloch, T.a. Costi, R.W. Helmes, D. Rasch, A. Rosch, Science **322**, 1520 (2008)
23. C. Schori, T. Stferle, H. Moritz, M. Khl, T. Esslinger, Phys. Rev. Lett. **93**, 240402 (2004), copyright (C) 2009 The American Physical Society
24. D. Clement, N. Fabbri, L. Fallani, C. Fort, M. Inguscio, Phys. Rev. Lett. **102**, 155301 (2009)
25. G.K. Campbell, J. Mun, M. Boyd, P. Medley, A.E. Leanhardt, L.G. Marcassa, D.E. Pritchard, W. Ketterle, Science **313**, 649 (2006)
26. A. Abanov, A.V. Chubukov, Phys. Rev. Lett. **84**, 5608 (2000)
27. M.A. Metlitski, S. Sachdev, Phys. Rev. B **82**, 075128 (2010)
28. A.J. Schofield, (2010), http://arxiv.org/abs/1001.4279
29. Q. Zhou, T.-L. Ho, Phys. Rev. Lett. **105**, 245702 (2010)

Chapter 14
Quantum Criticality: More Detailed Information

This chapter contains a number of short notes expanding details mentioned in the previous chapter.

14.1 Bose-Hubbard Model

The Bose-Hubbard model [1] describes cold bosonic atoms in an optical lattice [2]. It is defined by the Hamiltonian

$$H = -t\sum_{\langle i,j\rangle} b_i^\dagger b_j + \sum_i \left[\frac{U}{2} b_i^\dagger b_i^\dagger b_i b_i - \mu b_i^\dagger b_i\right] \tag{14.1}$$

where $\sum_{\langle i,j\rangle}$ indicates a sum over nearest neighbors i and j, and the operators b_i and $b_i^\dagger$ annihilate and create bosons on site i. They satisfy the canonical commutation relation $[b_i, b_j^\dagger] = \delta_{ij}$. The parameters t and U are controlled by the depth V_0 of the optical lattice defined by $V(r) = V_0(\cos(2\pi x/\ell) + \cdots)$. Energies are typically measured in terms of $E_R = \hbar^2\pi^2/(2m\ell^2)$. The values $t/U = 0.01$, 0.0585 shown in Fig. 13.3. correspond to lattice depths of $V_0/E_R = 18.1, 11.4$, respectively, for ^{87}Rb in a lattice with spacing $\ell = 532\,\mathrm{nm}$. This is found by numerically solving the non-interacting lattice problem and using the relations [2] $t = -\int d^3r w(\mathbf{r}-\mathbf{r_i})\left[-\frac{\hbar^2}{2m}\nabla^2 + V(\mathbf{r})\right] w(\mathbf{r}-\mathbf{r_j})$ and $U = \frac{4\pi\hbar^2}{m} a \int d^3r |w(\mathbf{r})|^4$, where $w(\mathbf{r})$ is the Wannier state, and $\mathbf{r_i}, \mathbf{r_j}$ are the locations of neighboring sites.

This chapter was adapted from "Techniques to measure quantum criticality in cold atoms" by by Kaden. R. A. Hazzard, Erich. J. Mueller, preprint available at arxiv:1006.0969 (2010).

K. R. A. Hazzard, *Quantum Phase Transitions in Cold Atoms and Low Temperature Solids*, Springer Theses, DOI: 10.1007/978-1-4419-8179-0_14,

14.2 Finite Temperature Gutzwiller Theory

We calculated the schematic finite temperature phase diagram in Fig. 13.1. within a finite temperature Gutzwiller approximation. Although only approximate this approach provides a qualitative picture of the role of temperature. More accurate phase diagrams can be obtained by using quantum Monte-Carlo techniques.

Following Fisher et al. [1] and Sachdev [3], we perform a Hubbard-Stratonovitch transform to decouple lattice sites: we introduce a new free field ϕ into the Bose-Hubbard Lagrangian and couple it to the b operators by terms $-\sum_i (b_i^\dagger \phi_i^\dagger + \text{H.c.})$. The new fields' Lagrangian is chosen so that upon integrating out ϕ the original Bose-Hubbard Hamiltonian is reproduced. This field ϕ can be interpreted as the order parameter of the superfluid state. We formally integrate out the b fields and expands the Lagrangian for the ϕ fields to quartic order in ϕ and second order in derivatives to obtain

$$\mathcal{L}_\phi = \mathcal{L}_0 + \phi^*(s + K_1\partial_\tau + K_2\partial_\tau^2 - \nabla^2)\phi + u/2|\phi|^4. \tag{14.2}$$

where the constants s, $\mathcal{L}_0$, K_1, K_2, and u can be expressed as integrals ($\mathcal{L}_0$ will give a non-universal contribution to the free energy considered in the next part). The mean field phase boundary is given by setting $s(T) = 0$. At $T = 0$ this reproduces the standard Gutzwiller theory. The explicit form for s is

$$s(T) = \frac{1}{zt} + \frac{1}{\mathcal{N}}\left[\sum_{m=0}^{\infty} \frac{m+1}{Um - \mu}\left(e^{-\beta(U/2)m(m-1)+\beta\mu m} - e^{-\beta(U/2)m(m+1)+\beta\mu(m+1)}\right)\right] \tag{14.3}$$

with $\mathcal{N} \equiv \sum_{m=0}^{\infty} e^{-\beta((U/2)m(m-1)-\mu m)}$. The sums converge rapidly, reducing the finding of the found boundary to the finding the solution of a nonlinear equation in one variable.

14.3 Non-Universal Contributions

In order to see the collapse described in the main text, one needs to subtract off the non-universal contributions to the density (n_0) and compressibility (κ_0). These non-universal contributions are readily calculated, or as described below can be extracted from numerical or empirical measurements.

For example, consider the Bose-Hubbard model, where the non-universal contributions arise from the constant term $\mathcal{L}_0$ in Eq. 14.2. This corresponds to the free energy density of decoupled sites, implying that the non-universal contribution to any on-site observable is found by evaluating the observable at $t = 0$ and at the U, T, and μ of interest. One finds

$$n_0 = \frac{\sum_{n=0}^{\infty} n e^{-\beta\epsilon_n}}{\sum_{n=0}^{\infty} e^{-\beta\epsilon_n}} \tag{14.4}$$

with

$$\epsilon_n = \frac{U}{2}n(n-1) - \mu n. \tag{14.5}$$

These sums converge quickly with n, and typically only a few terms are needed. In principle, one can also measure the function $n_0(T, \mu, U)$ by performing experiments at small t. In practice, there are a number of technical hurdles: equilibration is difficult in deep lattices [4], as is control and measurement of temperature [5], and for very deep lattices the Bose-Hubbard description breaks down [6–14].

As explained by Zhou and Ho [15], a more generic approach to accounting for the non-universal contributions is to expand about the quantum critical point. By construction n_0 is an analytic function, $n_0(\mu, T) = \bar{n} + \mathcal{O}(\mu - \mu_c, T)$, where $\bar{n} = n(\mu_c, 0)$ is a constant. Sufficiently close to the critical point, Eq. 14.3 simplifies to

$$T^{1-d/z}(\kappa - \bar{\kappa}) = \Upsilon\left(\frac{n - \bar{n}}{T^{d/z}}\right), \tag{14.6}$$

and the only remnant of the non-universal functions n_0 and κ_0 are the constants $\bar{\kappa}$ and $\bar{n}$. In the examples in this paper $\bar{\kappa} = 0$ and $\bar{n} = 0$ (vacuum to superfluid transition) or $\bar{n} = 1$ ($n = 1$ Mott to superfluid transition). In circumstances where these constants are nontrivial, they may be extracted by fitting experimental or numerical data. Zhou and Ho give a discussion of carrying out this expansion to higher order.

Using Eq. 14.6 we observe equally good collapse as we find from subtracting off the full n_0 in Eq. 14.4. In addition to its simplicity, the expression in Eq. 14.6 is completely in terms of simple constants and local observables. This can be contrasted with Eq. 14.4, where one needs the central chemical potential in the trap. Additionally, Eq. 14.6 is particularly powerful in that it applies even if one has only local equilibrium through the quantum critical regime.

14.4 Time of Flight Expansion

As an alternative to density profiles, one can study quantum criticality through time-of-flight expansion in which the trapping potential and interactions are turned off and the cloud is allowed to expand. At long times this maps the momentum distribution of the particles to the real space distribution. At low momenta, the system's behavior is fully universal. Kato et al. [16] show a representative calculated image of the momentum distribution in a quantum critical regime of the Bose-Hubbard model.

In typical cold atom experiments the inhomogeneous broadening from the trap makes it very difficult to extract quantum critical signatures from the expansion images: multiple regions of the trap contribute to the observed momentum

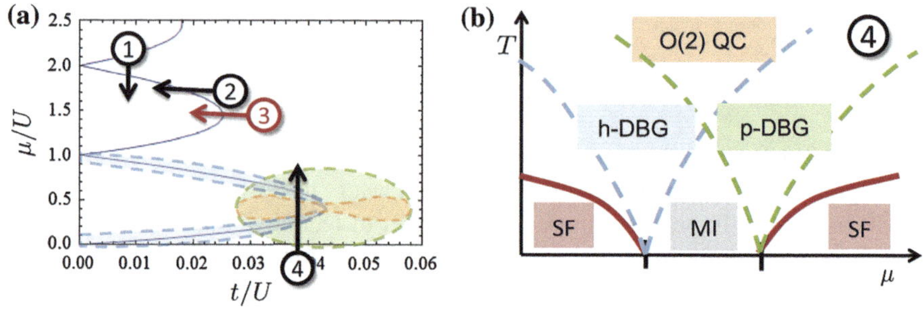

Fig. 14.1 (S1) Universality classes of the Bose-Hubbard model. **a** Annotated zero temperature Bose-Hubbard model phase diagram (cf. Fig. 13.1). Transitions via paths (1–2) are described by the dilute Bose gas (DBG) universality class while path (3) is described by the $O(2)$ universality class. *Shaded* regions schematically depict where each universality class holds: DBG physics governs the blue region along the lobe edge and $O(2)$ rotor physics governs the *orange* "bowtie" shaped region near the tip. The entire *green* region is described by the finite density $O(2)$ model. **b** Finite temperature phase diagram along path (4). At zero temperature, this path crosses two quantum phase transitions from a superfluid to a Mott insulator, and back. Each phase transition is in the DBG universality class and displays a quantum critical fan, inside of which the temperature sets the only length scale. $O(2)$ physics is recovered in the particle-hole symmetric regions where the quantum critical fans overlap. The "finite density $O(2)$" model universally governs this entire phase diagram including all of the crossovers (*dashed lines*) and phase transitions (*solid lines*) shown in this panel

distribution, including those far from the quantum critical regime (although the critical coupling value may be extracted, see e.g. Trotzky et al. [17] and Pollet et al. [18]). This difficulty can be circumvented by engineering a flat bottomed trapping potential. Such flat traps may also be advantageous for density probes of quantum criticality.

14.5 Finite Density $O(2)$ Model

This section describes more precisely the crossover between the dilute Bose gas and the $O(2)$ universality classes, and outlines open questions. Figure 14.1 illustrates the regions governed by each universality class. Along the Mott lobe edges, the physics is that of the dilute Bose gas (Fig. 14.1a, shaded blue regions). On the large-μ side of the Mott lobe, the relevant excitations are a dilute Bose gas of particles, while on the small-μ side, they are holes. Near the line of particle-hole symmetry passing through the tip of the lobe, both particles and holes are equally important and the physics is in the $O(2)$ universality class (Fig. 14.1a shaded orange region). Both of these universal structures are captured by the "finite density $O(2)$" model with imaginary time action $S = \int d^d r d\tau \mathcal{L}(\mathbf{r},\tau)$ defined by the Lagrangian

$$\mathcal{L} = -\phi^* \left[(\partial_\tau - \mu)^2 - c^2\nabla^2 + s \right] \phi + \frac{g}{2}|\phi|^4 \tag{14.7}$$

where ϕ is a complex bosonic field, evaluated at $\mathbf{r}, \tau$ in this expression. Here μ is the chemical potential, which controls the relative energy cost of holes and particles, s is the tuning parameter for the $\mu = 0$ phase transition, c the excitation velocity when $\mu = 0$, and g is the effective interaction strength. When μ is nonzero, at sufficiently low energies this reduces to the dilute Bose gas Lagrangian $\mathcal{L}_{\mathrm{DBG}} = \psi^*[\partial_\tau - \mu - \nabla^2/(2m)]\psi + (g/2)|\psi|^4$ (with ψ a simple rescaling of ϕ) since the quadratic time derivative is *irrelevant* in the renormalization group sense. When $\mu = 0$, this reduces to the $O(2)$ model defined by the Lagrangian $\mathcal{L}_{O(2)} = \phi^*(-\partial_\tau^2 - c^2\nabla^2 + s)\phi + (g/2)|\phi|^4$. The model defined by Eq. 14.7 predicts a scaling function [19–21]

$$n_u(\Delta_+, \Delta_-, T) = T^d \Psi_{O(2)+\mu}\left(\frac{\Delta_+}{T}, \frac{\Delta_-}{T}\right) \tag{14.8}$$

where $\Delta_\pm$ are the relevant gaps/superfluid stiffnesses at $\mathrm{T} = 0$. This scaling function is universal and reduces to the dilute Bose gas and $O(2)$ scaling in the appropriate limits. [To see this in the dilute Bose gas case, note that if for large Δ_-/T the scaling function goes to $\Psi(\Delta_+/T, \Delta_-/T) \to (\Delta_-/T)^{d/2}\Psi_r(\Delta_+/T)$ then $n_u \to T^{d/2}\Delta_-^{d/2}\Psi_r(\Delta_+/T)$. This reproduce the $z = 2$ scaling expected for the resulting dilute Bose gas case if $\Psi_n(\Delta_+/T) = \Delta_-^{d/2}\Psi_r(\Delta_+/T)$ is identified as the universal dilute Bose gas scaling function introduced in the main text.] This more general finite density $O(2)$ universal physics describes the entire the shaded green region of Fig. 14.1a.

A particularly insightful way to view the finite density *O(2)* crossovers is illustrated in Fig. 14.1b. This corresponds to a slice through Fig. 14.1a in the region described by Eq. 14.8. The low temperature behavior is dominated by the two dilute Bose gas quantum critical points, and the $O(2)$ physics emerges at higher temperatures where the two fans overlap. Not only does the scaling in Eq. 14.8 describe the physics in these fans, it also describes the low temperature phases.

14.6 Calculating Density Profiles of One-Dimensional Hardcore Bosons

For Fig. 13.2 we calculate our density profiles by mapping the one-dimensional, hardcore ($U \to \infty$), trapped lattice bosons described by the Hamiltonian

$$H = -t\sum_i \left[b_i^\dagger b_{i+1} + \mathrm{H.c.} + \frac{U}{2} b_i^\dagger b_i^\dagger b_i b_i + (V_i - \mu) b_i^\dagger b_i\right], \tag{14.9}$$

with harmonic trapping potential V_i, onto non-interacting fermions by the Jordan-Wigner transformation [22]

$$f_i \equiv \left[\prod_{j<i} (1 - 2b_j^\dagger b_j) \right] b_i. \tag{14.10}$$

Note that f_i and $f_i^\dagger$ satisfy the canonical anticommutation relation for fermions, $\{f_i, f_j^\dagger\} = \delta_{ij}$. This gives the non-interacting Fermi Hamiltonian

$$H = \sum_i \left[-t\left(f_{i+1}^\dagger f_i + \text{H.c.}\right) + (V_i - \mu) f_i^\dagger f_i \right]. \tag{14.11}$$

We numerically find the single particle eigenstates $\phi_i^{(\alpha)}$ with energy E_α. The bosonic density at site i is then equal to the fermionic density at site i by Eq. 14.10, and thus

$$n_i = \sum_\alpha \frac{1}{e^{\beta E_\alpha} + 1} |\phi_i^{(\alpha)}|^2. \tag{14.12}$$

14.7 Other Cold Atoms Systems Displaying Quantum Criticality

For reference, we provide a partial list of other quantum phase transitions in cold atoms systems. The most experimentally mature systems are: magnetic transitions in spinor gases [23], nematic transitions in dipolar gases [24], superfluid and magnetic transitions in partially polarized resonant Fermi gases [25], and transitions from fully polarized or fully paired phases to a partially polarized (FFLO) phase in one dimensional clouds of fermions [26]. There are also potentially a large variety transitions between of magnetic phases in multicomponent gases and mixtures [27]. Finally, there are ample opportunities to study the *trivial* dilute gas to vacuum transitions, governed by the chemical potential tuned zero temperature phase transition from a state with non-zero to zero density. This physics is found near the edge of every atomic cloud. Although well understood, these latter transitions are a good test of the analysis techniques.

Universal scaling behavior has already been experimentally studied in the two-dimensional dilute trapped Bose gas using techniques related to the ones we discuss here [28]. In the two dimensional Bose system which they were looking at, dimensional analysis alone suffices to provide the collapse: all of the irrelevant couplings are zero in the bare Hamiltonian describing this system.

14.8 Precise Definition of Universality

As we have emphasized, our procedure allows the extraction of *universal* scaling curves from non-universal observables such as density profiles. Here, we make precise the meaning of this universality and offer a fairly detailed derivation of the scaling forms. This summarizes some of Fisher et al. [1] scaling arguments in the present notation (see also Sachdev's book [3]). First we present the argument for a general observable. We additionally discuss the simplification that occurs when we specialize observables to the density and its derivatives, which take a simpler form due to the conservation of charge [20].

Scaling near a quantum critical point. First, we consider a general observable $\mathcal{O}(g_1,\ldots,g_n)$ that is a function of n coupling constants, chosen to vanish at the critical point. Using a coarse graining or renormalization group procedure, both $\mathcal{O}$ and the g_j's can be defined as functions of the scale ℓ. Near the fixed point of a renormalization group flow one generally has

$$g_j(\ell) = g_j \ell^{1/\nu_j} \tag{14.13}$$

as $\ell \to \infty$. The exponents ν_j define scaling dimensions for the corresponding couplings g_j, and g_j is defined as $g_j \equiv g_j(\ell = 1)$. Since the flow of coupling constants is defined to preserve the physics up to the scale,

$$\mathcal{O}(g_1(\ell),\ldots,g_n(\ell)) = \ell^{\alpha\mathcal{O}}\mathcal{O}(g_1,\ldots,g_n) \tag{14.14}$$

where $\alpha_{\mathcal{O}}$ is related to the scaling dimension of $\mathcal{O}$. We may solve Eq. 14.14 for $\mathcal{O}(g_1,\ldots,g_n)$ and choose ℓ such that $\ell = (A/g_n)^{\nu_n}$ for some constant A, yielding

$$\mathcal{O}(g_1,\ldots,g_n) = A^{-\alpha_{\mathcal{O}}\nu_n} g_n^{\alpha\mathcal{O}\nu_n} \mathcal{O}\left(\frac{g_1 A^{\nu_n/\nu_1}}{g_n^{\nu_n/\nu_1}},\ldots,\frac{g_{n-1}A^{\nu_n/\nu_{n-1}}}{g_n^{\nu_n/\nu_{n-1}}},A\right). \tag{14.15}$$

Thus, we see that the dependence of $\mathcal{O}$ on all the coupling constants is captured, up to a non-universal scale in both $\mathcal{O}$ and its arguments (determined by A), by a universal function. This is frequently abbreviated

$$\mathcal{O}(g_1,\ldots,g_n) = g_n^{\alpha_{\mathcal{O}}\nu_n}\Psi_{\mathcal{O}}\left(\frac{g_1}{g_n^{\nu_n/\nu_1}},\ldots,\frac{g_{n-1}}{g_n^{\nu_n/\nu_{n-1}}}\right) \tag{14.16}$$

with it implicitly understood that each argument and the overall factor has an associated non-universal scale. This is the sense in which our main text's equations should be understood as universal. We conventionally choose g_n to be the temperature T, obtaining

$$\mathcal{O}(g_1,\ldots,g_n) = A^{-\alpha_{\mathcal{O}}/z}T^{\alpha_{\mathcal{O}}/z}\mathcal{O}\left(\frac{g_1 A^{1/(z\nu_1)}}{T^{1/(z\nu_1)}},\ldots,\frac{g_{n-1}A^{1/(z\nu_{n-1})}}{T^{1/(z\nu_{n-1})}},A\right) \tag{14.17}$$

since $T \propto \xi^{-z} \propto \ell^z$ by the definition of the dynamical critical exponent z.

Hyperscaling. The scaling dimension of $\mathcal{O}$ is related to the scaling dimension $\nu_{\mathcal{O}}$ of the conjugate coupling constant $g_{\mathcal{O}}$, where $\mathcal{O} = \partial\mathcal{F}/\partial g_{\mathcal{O}}$. Under a renormalization group transformation, the free energy scales as $\mathcal{F} \propto \ell^{d+z}$ for universality classes satisfying hyperscaling (as is generally the case for those relevant to quantum criticality). Thus $\mathcal{O} \propto \ell^{d+z-1/\nu_{\mathcal{O}}}$, yielding $\alpha_{\mathcal{O}} = d + z - 1/\nu_{\mathcal{O}}$ and

$$\mathcal{O}(g_1,\ldots,g_n) = T^{1+d/z-1/(z\nu_{\mathcal{O}})}\mathcal{O}\left(\frac{g_1}{T^{1/(z\nu_1)}},\ldots,\frac{g_{n-1}}{T^{1/(z\nu_{n-1})}}\right), \tag{14.18}$$

where for notational clarity we have omitted the A's.

In the case of a conserved quantity, such as density, Sachdev [20] shows that one can relate $\nu_{\mathcal{O}}$ to z. Under an RG transformation, $n \propto \ell^d$, requiring $\alpha_n = d$. Thus $\nu_n = 1/z$ and

$$n_u = T^{d/z}\Psi_n\left(\frac{g_1}{T^{1/(z\nu_1)}},\ldots,\frac{g_{n-1}}{T^{1/(z\nu_{n-1})}}\right) \tag{14.19}$$

Zhou and Ho'e expression [15] emerges directly from Eq. 14.17 when μ is the only relevant coupling. As seen from Eq. 14.19, their expression simplifies to $n_u = T^{d/z}\Psi_u\left(\frac{\mu-\mu_c}{T}\right)$ regardless of the universality class of the phase transition.

14.9 Quantum Monte Carlo Parameters and Signal-to-Noise

We calculate the densities for the two-dimensional square lattice Bose-Hubbard model using worm algorithm quantum Monte Carlo algorithm [29] as implemented in the ALPS simulation package [30]. We performed a sufficient number of equilibration sweeps (10,000) and evaluation sweeps (30,000) to obtain accurate estimates of the density. Typical stochastic errors in the density were $\sim$0.1%, but these are amplified when we take derivatives to extract the compressibility. This stochastic error is comparable to imaging noise of radially averaged density profiles in real experiments [31] (1%). As seen in Fig. 13.3, our approach is robust against such noise. Even using the simplest finite difference approximations to derivatives—which greatly amplify noise—our technique provides accurate scaling curves up to noise levels of $\lesssim$1%. More sophisticated multi-point differentiation schemes are more robust to noise, and we expect these to be robust to noise of perhaps $\lesssim$5%. Even at the center of the cloud, some existing lattice experiments are already below this noise level and have demonstrated clean determination of both κ and n [31]. Furthermore, the noise level decreases with distance to the center due to trap averaging (although the chemical potential gradients eventually lead to breakdown of the Thomas-Fermi approximation, we find good agreement over a substantial portion of the cloud even, for example, for small of N $= 70$ 1D clouds). Systematic effects may be important as well, for example shot-to-shot total particle number fluctuations. These would have controlled, for example by post-selection of particle number, and this will moderately increase running time for the experiments.

Over most of the parameter space simulated, we find that systematic errors from the finite equilibration time are significantly smaller than the stochastic errors. We explore possible systematic errors using two methods: (1) running with longer equilibration times and (2) a jacknife binning analysis. The simulations were carried out for a system size of 10×10 lattice sites. Our results were insensitive to the finite size effects, except in the *classical* critical regime, mimicking the effects of a real trapping potential.

14.10 Finite-Size Scaling

We found by exact calculations of the trapped 1D hard core Bose-Hubbard model (and by mimicking the trapped 2D Bose-Hubbard model) that our procedure could extract the homogeneous system's universal scaling function within a $\lesssim 10\%$ accuracy assuming the Thomas-Fermi approximation holds, despite the finite particle number and trapping inhomogeneity. Nevertheless, it is possible to improve this further by accounting for the finite size scaling in the trap [15, 18, 32–34]. The trapping potential introduces a length scale L, and when this length is large compared to microscopic lengths, universal scaling in ξ/L^θ for some non-trivial θ results. For small values of this argument, the homogeneous system scaling is recovered. Accounting for this, some have explicitly demonstrated how to improve the accuracy of the naive Thomas-Fermi approach to determining critical coupling values (e.g., μ_c and T_c) [15, 18].

References

1. M.P.A. Fisher, P.B. Weichman, G. Grinstein, D.S. Fisher, Phys. Rev. B **40**, 546570 (1989)
2. D. Jaksch, C. Bruder, J.I. Cirac, C.W. Gardiner, P. Zoller, Phys. Rev. Lett. **81**, 3108 (1998)
3. S. Sachdev, *Quantum Phase Transitions* (Cambridge University Press, Cambridge, 2001)
4. C. Hung, X. Zhang, N. Gemelke, C. Chin, Phys. Rev. Lett. **104**, 160403 (2010)
5. D.M. Weld, P. Medley, H. Miyake, D. Hucul, D.E. Pritchard, W. Ketterle, Phys. Rev. Lett. **103**, 245301 (2009)
6. O.E. Alon, A.I. Streltsov, L.S. Cederbaum, Phys. Rev. Lett. **95**, 030405 (2005)
7. J. Li, Y. Yu, A.M. Dudarev, Q. Niu, New J. Phys. **8**, 154 (2006)
8. G. Mazzarella, S.M. Giampaolo, F. Illuminati, Phys. Rev. A **73**, 013625 (2006)
9. A. Smerzi, A. Trombettoni, Phys. Rev. A **68**, 023613 (2003)
10. P.R. Johnson, E. Tiesinga, J.V. Porto, C.J. Williams, New J. Phys. **11**, 093022 (2009)
11. J. Larson, A. Collin, J. Martikainen, Phys. Rev. A **79**, 033603 (2009)
12. K.R.A. Hazzard, E.J. Mueller, Phys. Rev. A **81**, 031602 (2010)
13. G.K. Campbell, J. Mun, M. Boyd, P. Medley, A.E. Leanhardt, L.G. Marcassa, D.E. Pritchard, W. Ketterle, Science **313**, 649 (2006)
14. S. Will, T. Best, U. Schneider, L. Hackermuller, D. Luhmann, I. Bloch Nature **465**, 197 (2010)
15. Q. Zhou, T.-L. Ho, Phys. Rev. Lett. **105**, 245702 (2010)
16. Y. Kato, Q. Zhou, N. Kawashima, N. Trivedi, Nat. Phys. **4**, 617 (2008)
17. S. Trotzky, L. Pollet, F. Gerbier, U. Schnorrberger, I. Bloch, N.V. Prokof'ev, B. Svistunov, M. Troyer, Nat. Phys. **6**, 998–1004 (2010). doi:10.1038/nphys1799
18. L. Pollet, N.V. Prokof'ev, B.V. Svistunov, Phys. Rev. Lett. **104**, 245705 (2010)
19. A.V. Chubukov, S. Sachdev, J. Ye, Phys. Rev. B **49**, 11919 (1994)
20. S. Sachdev, Zeitschrift für Physik B Condens. Matter **94**, 469 (1994)
21. S.A. Hartnoll, P.K. Kovtun, M. Muller, S. Sachdev, Phys. Rev. B **76**, 144502 (2007)

22. P. Jordan, E. Wigner, Zeitschrift für Physik A Hadrons Nuclei **47**, 631 (1928)
23. M. Ueda, Y. Kawaguchi, arxiv:1001.2072 (2010)
24. T. Lahaye, C. Menotti, L. Santos, M. Lewenstein, T. Pfau, Rep. Prog. Phys. **72**, 126401 (2009)
25. W. Ketterle, M.W. Zwierlein, arxiv arXiv:0801.2500v1 to appear in Ultracold Fermi Gases. in *Proceedings of the International School of Physics "Enrico Fermi", Course CLXIV, Varenna, 20–30 June 2006*, ed. by M. Inguscio, W. Ketterle (2008)
26. Y. an Liao, A.S.C. Rittner, T. Paprotta, W. Li, G.B. Partridge, R.G. Hulet, S.K. Baur, E.J. Mueller, 0912.0092 (2009)
27. I. Bloch, J. Dalibard, W. Zwerger, Rev. Mod. Phys. **80**, 885 (2008)
28. S.P. Rath, T. Yefsah, K.J. Günter, M. Cheneau, R. Desbuquois, M. Holzmann, W. Krauth, J. Dalibard, Phys. Rev. A **82**, 013609 (2010)
29. N.V. Prokof'ev, B.V. Svistunov, I.S. Tupitsyn, Phys. Lett. A **238**, 253 (1998)
30. A. Albuquerque, F. Alet, P. Corboz, P. Dayal, A. Feiguin, S. Fuchs, L. Gamper, E. Gull, S. Gürtler, A. Honecker, R. Igarashi, M. Körner, A. Kozhevnikov, A. Läuchli, S. Manmana, M. Matsumoto, I. McCulloch, F. Michel, R. Noack, G. Pawlowski, L. Pollet, T. Pruschke, U. Schollwöck, S. Todo, S. Trebst, M. Troyer, P. Werner, S. Wessel, J. Magn. Magn. Mater. **310**, 187 (2007)
31. N. Gemelke, X. Zhang, C. Hung, C. Chin, Nature **460**, 995 (2009)
32. M. Campostrini, E. Vicari, Phys. Rev. Lett. **102**, 240601 (2009)
33. M. Campostrini, E. Vicari, Phys. Rev. A **81**, 063614 (2010)
34. M. Campostrini, E. Vicari, Phys. Rev. A **81**, 023606 (2010)

Chapter 15
Systems Other than Cold Atoms

Although the bulk of this thesis concentrates on cold atomic gases, I have also worked in other areas of condensed matter physics. In graduate school, this has largely been in low temperature fluids and solids. The following chapters discuss some results I've obtained on these systems: collisional properties of nearly-degenerate spin-polarized hydrogen adsorbed on helium films (Chap. 16), anomalous behavior of atomic hydrogen defects in molecular hydrogen quantum solids (Chap. 17), and a possible scenario for supersolidity, inspired by experiments on solid hydrogen and helium (Chap. 18).

The systems studied here, especially the quantum solids, are in many ways the antithesis of cold atoms. In contrast to the simple, known, model Hamiltonians realized in cold atoms, in low temperature solids even the composition is frequently inhomogeneous, complicated, and unknown. This complexity opens up a richness of physics, but at the cost of being unsure if the basic models of the situation are correct. A theme in my work has been to try to give general constraints on possible theories to explain the phenomena, to give scenarios to explain behavior, and to make falsifiable predictions to test these theories.

K. R. A. Hazzard, *Quantum Phase Transitions in Cold Atoms and Low Temperature Solids*, Springer Theses, DOI: 10.1007/978-1-4419-8179-0_15,

Chapter 16
Film Mediated Interactions Alter Correlations and Spectral Shifts of Hydrogen Adsorbed on Helium Films

The work presented in this chapter gives a mechanism which is expected to reduce, potentially dramatically, the collision probability of two spin-polarized hydrogen atoms adsorbed on a helium film. This was motivated by experiments by Ahokas et al. which observed anomalously small collisional shifts. Refs. [1, 2] revealed that a symmetry missed in the original work is likely responsible for the reduction, however our mechanism is still relevant affects other observable properties: atomic recombination rates, other spectral lines, and suggests that the cloud will collapse at sufficiently low temperatures.

16.1 Chapter Overview

We argue that helium film-mediated hydrogen–hydrogen interactions dramatically reduce the magnitude of cold collision shifts in spin-polarized hydrogen adsorbed on a helium film. The magnitude of the reduction varies considerably across the possible range of experimental parameters, but this effect can consistently explain a two orders of magnitude discrepancy between previous theory and recent experiments. It can also qualitatively explain the anomalous dependence of the cold collision frequency shifts on the ^{3}He covering of the film. The mediated interaction is attractive, suggesting that the gas will become mechanically unstable before reaching the Kosterlitz-Thouless transition unless the experiment is performed in a drastically different regime, for example with a much thicker helium film.

This chapter was adapted from "Influence of Film-Mediated Interactions on the Microwave and Radio Frequency Spectrum of Spin-Polarized Hydrogen on Helium Film" by Kaden R. A. Hazzard and Erich J. Mueller, published in Physical Review Letters **101**, 165301 (2008).

K. R. A. Hazzard, *Quantum Phase Transitions in Cold Atoms and Low Temperature Solids*, Springer Theses, DOI: 10.1007/978-1-4419-8179-0_16,

16.2 Results

Two-dimensional (2D) dilute, spin polarized hydrogen is a quantum fluid that is actively studied by both the cold atom and low temperature communities. One driver of this interest is belief that 2D hydrogen can support a superfluid-normal Kosterlitz-Thouless transition, which crosses over to Bose-Einstein condensation (BEC) as the confinement length in the third-dimension is increased. The relatively large particle density and light mass give a high BEC temperature. Since the hydrogen–hydrogen interaction potential is accurately known, it is widely believed that one will be able to quantitatively compare theory and experiment.

Despite this quantitative understanding of dilute hydrogen, experiments on spin-polarized hydrogen adsorbed on a helium film display a cold collision frequency shift that is two orders of magnitude smaller than theory predicts [3]. This suggests that the helium, though traditionally considered inert, significantly affects the hydrogen. Here we show that a helium mediated interaction alters the hydrogen collision properties and resultantly reduces the cold collision shift. The exact magnitude of this shift depends on experimental parameters such as the helium film thickness. For some reasonable estimates of these parameters, we find that the magnitude of this reduction is sufficient to eliminate the discrepancy. Using these same estimates, we find a counterintuitive dependence of the shift on the ^{3}He concentration, just as seen in the experiments. We do not, however, find quantitative agreement with the shift's magnitude.

As elaborated below, Ahokas et al. [3] measure a two-photon radio frequency (rf)-microwave absorption spectrum. These experiments are in the "cold collision" regime [4], where the thermal de Broglie wavelength is longer than the interaction's effective range, resulting in extremely sharp spectral lines, which are shifted by interactions. These shifts are important: they limit the performance of technologies (hydrogen masers [5, 6], and atomic clocks [7]) and are versatile probes. For example, related spectral shifts were used to diagnose BEC in 3D spin polarized hydrogen [8], identify Mott insulating shells in optical lattices [9], and reveal pair structure in Fermi gases [10, 11]. Future experiments will rely on these techniques to explore ever more exotic phenomena.

Experimental details. Ahokas et al. [3] study a 2D hydrogen gas bound by van der Waals forces at a distance $\zeta \sim 5$ Å above a helium film, working in a uniform magnetic field $B = 4.6T$. They produce a cloud with more than 99% of the atoms initially in the lowest hyperfine state $|1\rangle$. Letting $|s_e s_n\rangle$ be the state with electron spin s_e and nuclear spin s_n, one finds $|1\rangle = \cos\theta|\downarrow\uparrow\rangle - \sin\theta|\uparrow\downarrow\rangle$, with $\theta = 5.5 \times 10^{-3}$. Driving a transition to state $|2\rangle = \cos\theta|\uparrow\downarrow\rangle + \sin\theta|\downarrow\uparrow\rangle$, the experimentalists measure spectra by counting the number of transferred atoms for a given probe detuning. They vary temperature, 2D hydrogen density, and the He film's ^{3}He concentration from negligible to 10^{14} cm $^{-2}$.

Calculating spectra. Neglecting, for now, the helium film's degrees of freedom, the hydrogen's Hamiltonian is

$$H = \sum_{j,\mathbf{k}} \epsilon_{j,\mathbf{k}} \psi^\dagger_{j,\mathbf{k}} \psi_{j,\mathbf{k}} + \sum_{\substack{i,j \\ \mathbf{k},\mathbf{p},\mathbf{q}}} \frac{V_{ij,q}}{2} \psi^\dagger_{j,\mathbf{p}} \psi^\dagger_{i,\mathbf{k}} \psi_{i,\mathbf{k}+\mathbf{q}} \psi_{j,\mathbf{p}-\mathbf{q}} \tag{16.1}$$

where $\psi^\dagger_{j,k}$ are bosonic creation operators for momentum $\hbar k$ and internal state j, $\epsilon_{j,k} = \hbar^2 k^2/2m + \delta_j - \mu_j$ is the free dispersion relation of the effectively 2D, mass m hydrogen, including the internal energy δ_j and chemical potential μ_j, $V_{ij,\mathbf{k}}$ is the Fourier space interaction potential between atoms in states i and j, and $\hbar$ is the reduced Planck's constant.

In the rotating wave approximation the probe's Hamiltonian is $H_P = \Omega_P \sum_{\mathbf{k}} e^{-i(\omega-(\mu_2-\mu_1))t} \psi^\dagger_{2,\mathbf{k}} \psi_{1,\mathbf{k}} + \text{H.c.}$ The rf and microwave photons transfer negligible momentum. Given that the range of the the potential (~1 Å) is significantly less than the 2D interparticle separation (~100 Å), the rf/microwave spectrum remains a delta function [4], with a shift given by [12]

$$\delta\omega = \frac{\hbar^2}{m} g_2(0)[f_{12}(q_T) - f_{11}(q_T)]\sigma_1, \tag{16.2}$$

where $g_2(r) \equiv \langle \psi_1^\dagger(r)\psi_1^\dagger(0)\psi_1(0)\psi_1(r)\rangle / \langle \psi_1^\dagger(0)\psi_1(0)\rangle^2$, $f_{ij}(q_T)$ is the i–j scattering amplitude evaluated at the system's characteristic momenta $\hbar q_T$, and $\sigma_1 \equiv \langle \psi_1^\dagger(0)\psi_1(0)\rangle$ is the 2D density. Here $q_T = \sqrt{m k_B T}/\hbar$ is proportional to the thermal momentum, where k_B is Boltzmann's constant. In a dilute thermal gas $g_2(0) = 2$. A similar formula holds in 3D, where $f_{ij} = 4\pi a_{ij}$ with a_{ij} the 3D scattering length.

In quasi-2D, where kinematics are 2D but the 3D scattering length a is much less than the perpendicular confinement length, we can construct a 2D interaction that reproduces the low-energy scattering properties. Assuming harmonic confinement with oscillation frequency ω_{osc}, the effective 2D scattering amplitude is [13, 14]

$$f = 2\sqrt{2\pi} \frac{\hbar^2}{l/a - (1/\sqrt{2\pi}) \log(\pi q_T^2 l^2)} \tag{16.3}$$

where $l = \sqrt{\hbar/(m\omega_{\text{osc}})}$ is the length scale of z-axis confinement of the hydrogen gas. Following Ref. [3]'s discussion, the characteristic length scale for the confinement in their experiments is $l_0 = \sqrt{2\pi} l = \hbar/\sqrt{2mE_a} \sim 5$ Å where E_a is the adsorption energy of the hydrogen on the helium film. The spectral shift is then

$$\delta\omega = \frac{4\pi\hbar^2}{m} \left[\frac{1}{l_0/a_{12} - \log(q_T^2 l_0^2/2)} - \frac{1}{l_0/a_{11} - \log(q_T^2 l_0^2/2)} \right] g_2(0)\sigma_1. \tag{16.4}$$

If $\{a_{11}, a_{12}\} \ll l_0$ this reduces to the simpler form

$$\delta\omega = \frac{4\pi\hbar^2}{m} g_2(0)(a_{12} - a_{11})\frac{\sigma_1}{l_0}, \tag{16.5}$$

as for a 3D gas with density σ_1/l_0.

The spectral shift predicted by Eq. 16.5 is one hundred times larger than experiment. We will show that this discrepancy is consistent with the corrections found by including the hydrogen–hydrogen interaction mediated by the helium film, V_{med}. Although V_{med} is state-independent, it alters $\delta\omega$ by reducing the probability that two particles are close enough to feel the spin dependent V_{ij}. Remarkably, even an attractive V_{med} can reduce this overlap [15].

Helium film-mediated interaction. Wilson and Kumar derive a hydrogen-only effective action [16] assuming that (i) the helium film's excitations are non-interacting, (ii) the hydrogen–helium interaction potential is modeled as a van der Waals potential, $V_{\text{H-He}}(r) = -(6\Lambda_0/n\pi)r^{-6}$ with Λ_0 controlling the potential's strength, n is the helium density, and r is the 3D distance between the hydrogen and helium atoms, (iii) the hydrogen confinement length l_0 is significantly smaller than the hydrogen-helium separation ζ, and (iv) retardation effects can be neglected. We define the total hydrogen density operator $\rho_t(\rho) \equiv \sum_j \psi_j^\dagger(\rho)\psi_j(\rho)$ where ρ is the atomic coordinate projected onto the plane of the 2D film. Then Fourier transforming Ref. [16]'s effective Hamiltonian gives

$$H_{\text{eff}} = H - \frac{1}{2}\int\int d^2\rho d^2\rho' \rho_t(\rho)\rho_t(\rho')V_{\text{med}}(\rho - \rho') \tag{16.6}$$

with mediated pair interaction

$$V_{\text{med}}(\mathbf{R}) = V_0 \bar{V}_{\lambda/\zeta}(\mathbf{R}/\lambda), \tag{16.7}$$

$$V_0 \equiv \frac{2\delta^2}{\pi^3\lambda^2 M C_3^2} \tag{16.8}$$

defining $\delta \equiv 6\Lambda_0\phi_g/n\zeta^4$, $\lambda \equiv \sqrt{\frac{1}{MC_3^2}\left(\frac{1}{2M} + \frac{\beta d_0}{n}\right)}$, and $\phi_g \equiv \sqrt{n d_0}$, where C_3 is the film's third sound speed, d_0 is the helium film thickness, M is the helium mass, and β is the film surface tension (estimates of parameters in experiments are given later). The non-dimensionalized potential $\bar{V}_\xi$ is found to be

$$\bar{V}_\xi(x) = \int\int d^2\bar{\rho} d^2\bar{\rho}' \bar{A}(\bar{\rho})\bar{A}(\bar{\rho}')\bar{G}_F\left(\frac{\mathbf{x} + \rho + \rho'}{\xi}\right), \tag{16.9}$$

where ξ is an argument of V_ξ, and we define

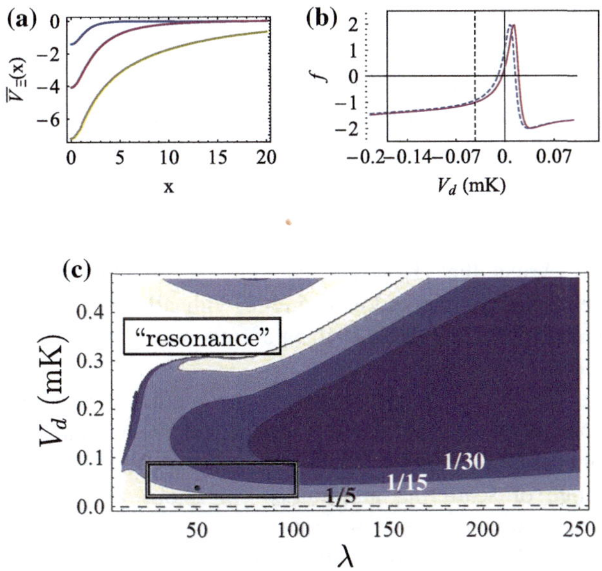

Fig. 16.1 **a** Rescaled mediated potential as a function of interparticle separation divided by λ, as given by Eq. 16.9, with $\xi = 1, 4, 15$ from top to bottom. **b** Scattering amplitudes f as a function of mediated potential depth factor $V_d \equiv (2\delta^2)/(\pi^3 \lambda_0^2 M C_3^2) = V_0(\lambda/\lambda_0)^2$ with $\lambda_0 \equiv 50$ Å, for the triplet (*dashed*) and singlet (*solid*) scattering channels. Vertical lines indicate V_d for the typical parameters given in the text (*dashed line*) and for $V_d = 0$ (*solid*). The divergences near $V_d = 0$ have been rounded off for display. **c** Contour plot of the factor by which V_{med} reduces the frequency shift as a function of λ and V_d. Typical values of parameters are shown by the black dot, while factor of two variations comprise the interior of the black rectangle. The contour plot is hidden in the white "resonance" region (far away from the relevant regime), where the scattering amplitude diverges

$$\bar{A}(x) \equiv \frac{1}{(1+x^2)^3} \quad \text{and} \quad \bar{G}_F(x) \equiv K_0(x), \tag{16.10}$$

with K_0 the zeroth modified Bessel function of the second kind. Note that $A(\rho) = -(12\phi_g \Lambda_0/\pi n \zeta^6)\bar{A}(\rho/\zeta)$ is the hydrogen–helium van der Waals interaction at a separation ρ and $G_F(\rho) = (n/2d_0\pi\beta)\bar{G}_F(\rho/\lambda)$ is the helium film's Green's function. We numerically compute $\bar{V}_\xi(x)$ as a function of x and ξ. Typical results are shown in Fig. 16.1a.

Spectral line shifts with mediated potential. To evaluate spectral shifts via Eq. 16.2, we calculate the scattering amplitude of $V_{\text{tot}} = V_{ij} + V_{\text{med}}$. Since the range of the bare hydrogen-hydrogen potential r_{eff} is much smaller than the range of V_{med}, we may replace V_{tot} with $V'_{\text{tot}} = V'_{ij} + V_{\text{med}}$, where V'_{ij} is an arbitrary short range potential reproducing V_{ij}'s scattering amplitude. This replacement requires that the relative momentum $k_0 \sim \sqrt{|2mV_{\text{med}}(0)|}/\hbar$ of the particles when they reach the bottom of the attractive potential V_{med} satisfies $k_0 r_{\text{eff}} \ll 1$, which is well-satisfied.

The same conditions allow us to use a boundary condition in place of V'_{ij}, chosen to reproduce V'_{ij}'s scattering amplitude [15]. We numerically solve the resulting two-particle Schrödinger equation with potential V'_{tot}.

We use the following estimates in our calculations, taken from Refs. [3, 16, 17]: $l_0 \sim 5\,\text{Å}, a_t = 0.72\,\text{Å}, a_s = 0.17\,\text{Å}, \lambda \sim 50\,\text{Å}, C_3 \sim 1\text{m/s}, q_T \sim (30\,\text{Å})^{-1}$, $\delta \sim 0.265\sqrt{\hbar^2 MC_3^2/2m}$, and $\zeta \sim 5\,\text{Å}$. These are typically known within a factor of two.

Figure 16.1b shows the $|1\rangle - |1\rangle$ and $|1\rangle - |2\rangle$ scattering amplitudes as a function of the mediated interaction strength, with all other parameters fixed at their typical values. The two vertical lines correspond to zero mediated potential and typical mediated potential strengths.

Figure 16.1c shows the factor by which the mediated potential reduces the spectral shift $\delta\omega$, relative to $\delta\omega$ in the absence of the mediated potential, as a function of the potential depth and the characteristic length λ. The black box denotes the range of scattering amplitudes consistent with known parameters. A black dot indicates our best estimate of typical experimental parameters. Figure 16.1c shows that for our best estimate of parameters, the mediated potential reduces $\delta\omega$ by a factor of 7, while a nearly 30-fold decrease is possible within the range of experimental parameters. The reductions we quote are due solely to the mediated potential. Compared with the analysis in [3], there is an additional 40% reduction of $\delta\omega$ simply by using the more accurate formula of Eq. 16.2 in place of the approximation Eq. 16.5. Thus the observed reduction in the cold collision frequency shift is consistent with our predictions.

^{3}He film—Our theory also accounts for unexpected effects of adding ^{3}He to the film's surface. Since adding ^{3}He reduces the hydrogen surface adsorption energy, one would naively expect an increase in the confinement length, decrease in density, and decrease in spectral shift. Instead, adding ^{3}He with surface density $\sim 10^{14}$ cm^{-2} leads to an observed 25% *increase* in the shift. Our theory predicts such an increase, but with larger magnitude.

The ^{3}He driven change in the adsorption energy E_a modifies the mediated interaction by increasing the confinement length l by $\sim 2 - 10\%$ and by increasing ζ, the distance of the gas to the helium surface. The fractional change in ζ should be comparable to the fractional change in l, and we take these to be equal. We find a 200–400% increase of frequency shifts upon adding ^{3}He.

This reduction can be understood by the following simple argument, which focuses on the dependence of V_0 on l. Eq. 16.8 shows that the mediated potential roughly scales as $V_{\text{med}} \propto 1/\zeta^8 \propto 1/l^8$, so a 2–10% increase of l yields a $\sim$15–53% decrease of the mediated potential strength, moving the dashed line in Fig. 16.1b to the right by 15–53%, leading to a drastic increase in the shift. This simple argument omits the $1/l$ factor in $\delta\omega$, due to the decreased 3D density, and $\bar{V}_{\lambda/\zeta}$'s dependence on l, but as our more detailed calculations show these effects are insufficient to compensate.

Validity of Approximations. We have made a number of approximations. Here we enumerate the most important ones, and discuss which ones need to be addressed in the future through a more sophisticated theory.

The most severe approximation we made is to follow Ref. [16], and neglect retardation in the induced potential. Such effects are relevant when the phase speed of the hydrogen excitations ω/k becomes large compared to C_3 at characteristic speeds $\sqrt{k_B T/m}$ and energies $k_B T$. For $T \sim 50$ mK as in the experiments, $\omega/k \sim 5$ m/s. In comparison, typical third sound speeds are $C_3 \sim 1$ m/s, so we expect retardation corrections may be significant. Including the frequency dependence in calculating the spectral shift is challenging, requiring solution of a coupled set of 2D partial differential equations for each ω. We expect that a sharp spectral peak survives, but with reduced spectral weight.

Additionally, given the long length-scale of the mediated potential, we should also critically examine the assumption that only the long-wavelength limit of the s-wave phase shift is needed to evaluate the cold collision frequency shift. Eq. 16.2 requires that the areal interparticle distance ($n_{2D}^{-1/2}$) is larger than the effective range of the potential R_e. Since $R_e \sim 250$ Å and $n_{2D}^{-1/2} \sim 100-300$ Å, the approximation with only the $k = q_T$ s-wave scattering shift may not be sufficiently accurate. Similarly, the thermal wavelength is $\lambda_T \sim 30$ Å while the cold collision regime strictly requires $\lambda_T \gg R_e$. This can smear out spectral lines somewhat, although since $\lambda_T \sim R_e$ and the potential is rather shallow, the spectral lines may remain quite sharp. Here retardation helps us, as the slower moving atoms are the dominant contributors to the spectral peak, and these have a substantially larger de Broglie wavelength—a factor of 5 for the fastest contributing atoms.

Finally, we have neglected mass renormalization coming from virtual desorption–adsorption processes. This approximation is well-justified since the desorption process is relatively slow [18, 19], implying that the mass is renormalized by at most a few percent [17].

Conclusions. In summary, we have shown that incorporating the helium surface into the theory of the hydrogen gas provides a significant renormalization of spectral shifts, in addition to a $\sim 40\%$ reduction from previous estimates by properly treating the quasi-2D nature of the hydrogen. For a reasonable parameter range these shifts are consistent with experimental observations. In addition, we have shown that adding ^{3}He to the helium film increases spectral shifts, in agreement with experiment and in contrast to the naive theory.

While the mediated interaction provides a consistent explanation of the experimental observations, it is likely that it is not the whole story. A 3D control experiment finds an ill-understood reduction of $\delta\omega$ relative to theory in the 3D gas (S. Vasiliev, private communication). (Note: since the publication of the original paper, this control experiment has been invalidated.) The same physics is likely to be also playing a role in 2D. Given the known uncertainty in the experimental parameters, it would be quite reasonable for the mediated interaction to be responsible for only a fraction of the observed shift.

We have found that the helium surface induces an *attractive* interaction between the hydrogen atoms. It is useful to speculate on what other physical effects this interaction can cause. For example, can it drive a mechanical instability? While thermal pressure will stabilize the gas at the temperatures currently being studied, we believe that as the temperature is lowered that a "collapse" might occur, similar to the ones seen in atomic gases [20–22]. Importantly, the Kosterlitz-Thouless transition requires repulsive interactions, so the mediated interaction may eliminate the possibility of achieving superfluidity without significantly altering experimental parameters, for example by increasing the helium film thickness.

Finally, we point out possible ramifications of our theory to ultracold atomic gases and elsewhere. The key to our findings is that in quasi-2D, infinitesimal attractive interactions generate zero energy scattering resonances. These are pushed to finite positive energy by the 3D interaction, but a genuine 2D interaction easily overwhelms the effects of the 3D interaction. One can imagine our present theory applying to quasi-2D fermi-bose or bose–bose mixtures where one system mediates an interaction for the other species. Interestingly, similar physics might arise in layered systems: one layer effectively mediates an interaction in adjacent layers—incorporating these effects in a consistent manner could lead to dramatically modified interaction properties for each layer.

Acknowledgments KH thanks Neil Ashcroft, Sourish Basu, Stefan Baur, Dan Goldbaum, Jarno Järvinen, Joern Kupferschmidt, and Sophie Rittner for conversations, and especially Sergei Vasiliev and Dave Lee for an introduction to Ref. [3]'s experiments. This material is based upon work supported by the National Science Foundation through grant No. PHY-0758104.

References

1. J. Ahokas, J. Järvinen, G.V. Shlyapnikov, S. Vasiliev, Phys. Rev. Lett. **101**, 263003 (2008)
2. A.I. Safonov, I.I. Safonova, I.I. Lukashevich, JETP Lett. **87**, 23 (2008)
3. J. Ahokas, J. Järvinen, S. Vasiliev, Phys. Rev. Lett. **98**, 043004 (2007)
4. C.J. Pethick, H. Smith, *Bose-Einstein condensation in dilute gases*. (Cambridge University Press, The Edinburgh BuildingCambridge CB2 2RU, 2001)
5. R.L. Walsworth, I.F. Silvera, E.M. Mattison, R.F.C. Vessot, Phys. Rev. A **46**, 24952512 (1992)
6. M.E. Hayden, M.D. Hürlimann, W.N. Hardy, Phys. Rev. A **53**, 15891604 (1996)
7. K. Gibble, S. Chu, Phys. Rev. Lett. **70**, 1771 (1993)
8. D.G. Fried, T.C. Killian, L. Willmann, D. Landhuis, S.C. Moss, D. Kleppner, T.J. Greytak, Phys. Rev. Lett. **81**, 3811 (1998)
9. G.K. Campbell, J. Mun, M. Boyd, P. Medley, A.E. Leanhardt, L.G. Marcassa, D.E. Pritchard, W. Ketterle, Science **313**, 649 (2006)
10. C.H. Schunck, Y. Shin, A. Schirotzek, M.W. Zwierlein, W. Ketterle, Science **316**, 867 (2007)
11. Y. Shin, C.H. Schunck, A. Schirotzek, W. Ketterle, Phys. Rev. Lett. **99**, 090403 (2007)
12. M. Öktel, T.C. Killian, D. Kleppner, L.S. Levitov, Phys. Rev. A **65**, 033617 (2002)
13. D. Petrov, G. Shlyapnikov, Phys. Rev. Lett. **84**, 2551 (2000)
14. D. Petrov, D. Gangardt, G. Shlyapnikov, J. Phys. IV **116**, 5 (2004)
15. L. Landau, E. Lifshitz, *Statistical Physics I*. (Butterworth Heinemann, London, 1997)
16. B.G. Wilson, P. Kumar, Phys. Rev. B **27**, 3076 (1983)

17. J.P.H.W.v.d Eijnde, C.J. Reuver, B.J. Verhaar, Phys. Rev. B **28**, 63096315 (1983)
18. J.J. Berkhout, J.T.M. Walraven, Phys. Rev. B **47**, 8886 (1993)
19. B.W. Statt, Phys. Rev. B **32**, 7160 (1985)
20. C.A. Sackett, H.T.C. Stoof, R.G. Hulet, Phys. Rev. Lett. **80**, 2031 (1998)
21. E.J. Mueller, G. Baym, Phys. Rev. A **62**, 053605 (2000)
22. J.L. Roberts, N.R. Claussen, S.L. Cornish, E.A. Donley, E.A. Cornell, C.E. Wieman, Phys. Rev. Lett. **86**, 4211 (2001)

Chapter 17
Candidate Theories to Explain the Anomalous Spectroscopic Signatures of Atomic H in Molecular H_2 Crystals

17.1 Introduction and Motivation

Quantum solids, where the zero point motion of the atoms is greater than roughly 10% of their separation, form a fascinating class of materials. A principal question with these materials is to what extent they are quantum coherent, and under what conditions they can be supersolid—supporting dissipationless mass flow. Examples of quantum solids include ^{4}He [1–5], solid hydrogen [6], Wigner crystals, and atomic hydrogen defects in solid molecular hydrogen [7–9]. Here we theoretically study the last system, giving a critical evaluation of scenarios of Bose-Einstein condensation of atomic hydrogen defects. We make testable predictions for these scenarios.

Solid hydrogen, with a Lindemann ratio of 0.18, is the only observed molecular quantum crystal. It is a rich system with many rotational order/disorder transitions [10]. The phenomenology of H_2 solids is even more interesting when atomic H defects are introduced [7–9]. Here we focus on the spectroscopic properties of this system, and how they may be related to quantum coherence.

Our work is motivated by recent experiments in low-temperature ($T \sim 150\,\mathrm{mK}$) solid molecular hydrogen, populated with large densities ($n \sim 10^{18}\,\mathrm{cm}^{-3}$) of atomic hydrogen defects. These experiments observe unexplained internal state populations [7–9]. Ahokas et al. [7] provocatively conjectured that the anomalies may be related to Bose-Einstein Condensation (BEC) of the atomic defects. Here our goal is to explore and constrain this and alternative scenarios. We conclude that global Bose-Einstein condensation could not realistically explain the experiments, and that local condensation would lead to distinct signatures in future experiments.

This chapter was adapted from "Atomic H in molecular{H}_2crystals: candidate theories of experimental anomalies" by Kaden. R. A. Hazzard, Erich. J. Mueller, Physical. Review. B 82, 014303 (2010) [selected as an "Editor's Suggestion"].

K. R. A. Hazzard, *Quantum Phase Transitions in Cold Atoms and Low Temperature Solids*, Springer Theses, DOI: 10.1007/978-1-4419-8179-0_17,

17.2 Experiments

We review Ahokas et al.'s [7] experimental apparatus, results, and observed anomalies.

17.2.1 General Introduction: Physics of Atomic Hydrogen Embedded in Solid Hydrogen

Hyperfine structure of atomic hydrogen. Figure 17.1 schematically shows the level structure for a H atom in a $B = 4.6\,\mathrm{T}$ magnetic field, similar to that used in the experiments of interest [7]. At these large fields, the levels break into two nearly degenerate pairs. Levels within a pair are separated by radio frequencies and the pairs are separated by microwave frequencies. The electronic spin in states a and b is aligned with the magnetic field and is anti-aligned in the other states.

Ahokas et al. [7] observe that within the solid hydrogen matrix the atomic hydrogen's spectra is modified. The a-d energy splitting decreases while the b-c energy splitting increases by the same amount.

Spectroscopy of atomic hydrogen. We consider three different spectroscopic probes, distinguished by the states they connect: nuclear magnetic resonance (NMR), electron spin resonance (ESR), and electron-nuclear double resonance (ENDOR). NMR uses radio waves to couple the$a$ and b states, while ESR uses microwaves to couple the a and d or b and c states. ENDOR uses a two photon transition to couple a and c. All of these probes can be used in the linear regime, where they give information about the splittings and occupation numbers, or in the non-linear regime. As an example of the latter, Ahokas et al. [7] study what happens when they apply high RF power, saturating the NMR line.

Molecular hydrogen. At these temperatures, two states of molecular hydrogen are relevant: the "para" and "ortho" states. In the "para" or "p-" configuration the relative nuclear wavefunction ψ is symmetric under exchanging nuclei and the nuclear spins form a singlet. In the "ortho" or "o-" configuration ψ is antisymmetric and the nuclear spins form a triplet. In all cases, the electrons are in a symmetric bonding orbital and the electronic spins are consequently anti-aligned.

The true ground state of solid hydrogen is formed of p-H_2. In practice, however, it is very rare to have a pure p-H_2 sample. The ortho state is long lived [10], requiring hours for the concentration to change by 1%. This ortho-para conversion can be a source of heating in experiments with an energy $\Delta/k_B = 170\,\mathrm{K}$ released per molecule. In the experiments of Ahokas et al. [7] the exact quantity of o-H_2 is unknown, but given the growth technique it is likely to be at least ten percent.

At standard pressure the p-H_2 in the solid is highly spherical: interaction with neighboring H_2 negligibly distorts the p-H_2. Modeling the hydrogen-hydrogen interactions by their vacuum values quite accurately describes quantities such as the speed of sound [10]. If sufficient ortho-hydrogen is present, orientational

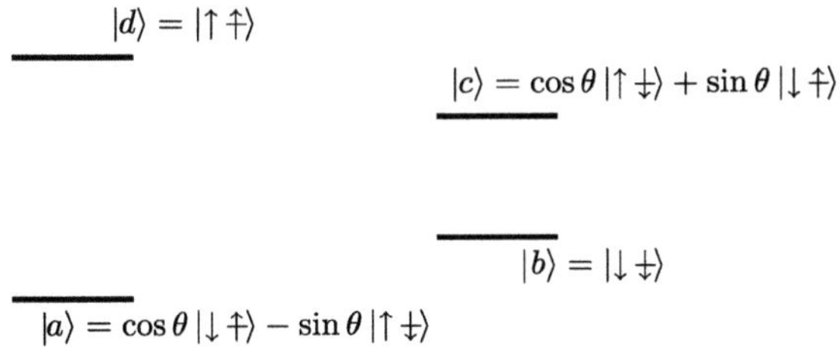

Fig. 17.1 Hyperfine level diagram for hydrogen atom in a $B = 4.6\,\mathrm{T}$ magnetic field, where the mixing angle is $\theta = 3 \times 10^{-3}$. Vertical axis schematically denotes energy, while horizontal axis has no physical meaning. *Arrows* denote electron (*no-slash arrow*) and nuclear (*slashed arrow*) spin projections

ordering transitions may occur around 1 K. The models we consider do not rely upon any orientational ordering. Depending on sample-preparation conditions, either hcp or fcc crystals may be produced [10].

Atomic hydrogen in the solid lattice. The motion of atomic hydrogen in molecular hydrogen has been widely studied [6]. Both thermally activated and quantum tunneling contribute to defect motion, but quantum tunneling dominates at these low temperatures (the two dominant tunneling pathways have energy barriers of 4600 K and 100 K). One motivation for these studies is to explore the possibility of Bose-Einstein condensation of these defects. This is conceptually related to supersolidity driven by condensation of vacancies.

Kumada [6] argues on the basis of experimental data that the exchange reaction $H + H_2 \rightarrow H_2 + H$ is the dominant diffusion mechanism at low temperatures. Other tunneling pathways are possible, including correlated, collective relaxation and "physical" diffusion.

Ahokas et al. [7] achieve populations of 50 ppm H defects in their solid, and argue that H sits at substitutional sites. The observed lifetime of these defects was weeks. The dominant decay mechanism should be the recombination of two hydrogen defects. One therefore expects that this rate is determined by the diffusion rate of the defects. The long lifetime is therefore inconsistent with the diffusion rates predicted by phonon assisted tunneling. One possible explanation is the suppression of tunneling by the strain-induced mismatch of energy levels between neighboring sites [11, 12].

Crystal growth Ahokas et al. [7] grow solid hydrogen from a gas of electron-spin polarized metastable hydrogen atoms, which undergo two-hydrogen recombination to form molecules—these events are only allowed in the presence of walls. The H_2 solid grows layer-by-layer from at a rate of 0.5–1 molecular layer per hour. After ~ 1 week, a quartz microbalance revealed a film thickness of 150 ± 1 layers.[1]

[1] private communication with Jarno Järvinen

17.2.2 Anomalies and Experimental Results

Ahokas et al. [7] observed four anomalies: (1) several orders-of-magnitude too fast "Overhauser" relaxation, (2) a non-Boltzmann a-b population ratio, (3) saturation of the a-b spectroscopic line fails to give a 1:1 population ratio, and (4) recombination rates are extremely low. Items 2 and 3 will be our main focus.

Overhauser relaxation. The c to a relaxation is expected to be extremely small at the 4.6 T fields of the experiments. This can be seen from the small mixing angle $\theta \approx 3 \times 10^{-3}$, indicated in Fig. 17.1. The mixing angle appears in the states as $|a\rangle = \cos\theta|\downarrow \Uparrow\rangle - \sin\theta|\uparrow \Downarrow\rangle$, and $|c\rangle = \cos\theta|\uparrow \Downarrow\rangle + \sin\theta|\downarrow \Uparrow\rangle$. Any c-a decay mechanism by photon emission is suppressed by θ^{α} with [13] $\alpha \gtrsim 1$. Ahokas et al. [7] observe no such suppression: the c-a line decays with a time constant of $\lesssim 5$ s, similar to the d-a decay time.

Equilibrium populations. The polarization

$$p \equiv \frac{n_a - n_b}{n_a + n_b} \tag{17.1}$$

characterizes the a and b state population. Assuming a Boltzmann distribution $n_a/n_b = \exp(\Delta_{ab}/T)$ where $\Delta_{ab} \approx 43$ mK is the difference in energies between the b and a states, one expects $p = 0.14$ at 150 mK. On the contrary, Ahokas et al. [7] measure $p = 0.5$.

When one of the states is depleted the system returns to this non-Boltzmann value on a time scale of ~ 50 hours.

Saturation of a-b line. An extremely strong rf field should saturate the a:b line, driving the population ratio to 1:1, or $p = 0$. Ahokas et al. [7] obtained a minimum of $p = 0.2$ at high excitation powers. For sufficiently large power this saturated population ratio was independent of the applied rf power.

Low recombination rates. As previously described, at 150 mK Ahokas et al.'s [7] recombination rate is much smaller than expected, negligible on a time scale of weeks. In contrast, at $T = 1$ K their recombination rates are consistent with previous studies [14].

Hole burning. Ahokas et al. applied a magnetic field gradient and a microwave field to saturate the a-b line (depopulating b atoms to the c state, which then relax to a state via the "forbidden" c-a transition) in a millimeter sized region of the sample. The spectral hole recovered in a time similar to that in the homogeneous case, indicating that the nuclear spin-relaxation is somewhat faster than spin migration. This would seem to indicate that the atomic hydrogen defects are immobile on regions much larger than a millimeter.

17.3 Scenarios

In this section we evaluate some previously proposed scenarios for these phenomena and suggest a new one [7]. For each scenario, we present the idea, examine its consistency with Ahokas et al.'s [7] experiments, consider possible microscopic mechanisms, and *give testable predictions.*

17.3.1 Bose Statistics and Bose-Einstein Condensation

17.3.1.1 Idea

Ahokas et al. [7] suggested Bose-Einstein condensation (BEC) as a possible mechanism to explain the departure from the Boltzmann distribution. In a BEC the lowest energy mode becomes "macroscopically occupied," leading to an excess of a-H. Neglecting interactions, the BEC transition temperature T_c for a homogeneous system of spinless particles with density ρ and effective mass m^* is

$$T_c = \left(\frac{\rho}{\zeta(3/2)}\right)^{2/3} \frac{2\pi\hbar^2}{k_B m^*}. \tag{17.2}$$

Here ζ is the Riemann zeta function; $\zeta(3/2) \approx 2.61$ [15]. Including the b states is straightforward and makes only small changes: for example if $\Delta_{ab} = 0$ one would divide the density by a factor of 2. Note that as m^* becomes larger the transition temperature becomes smaller.

17.3.1.2 Phenomena Explainable

This scenario can in principle explain the non-Boltzmann equilibrium ratio n_b/n_a. It does not provide an explanation of the inability to saturate the a–b line, the slow recombination, or the fast Overhauser cross-relaxation.

17.3.1.3 Consistency with Experiment

Transition temperature and densities. As reported in Ahokas et al. [7], for density $\rho \sim 10^{18}\text{cm}^{-3}$ and effective mass m^* similar to bare mass m, the ideal Bose gas transition temperature in (17.2) is $T_c \sim 30\,\text{mK}$, far below the experimental temperature. To address this inconsistency Ahokas et al. [7] suggest that phase separation may concentrate the defects to locally higher densities. For example, if the defects phase separated so that their density was $\rho \sim 3 \times 10^{19}\,\text{cm}^{-3}$, then with $m^* = m$ the transition temperature would be $T_c = 170\,\text{mK}$, and one would reproduce the observed ratio of n_a/n_b at $T = 150\,\text{mK}$. Such a powerful concentrating mechanism would have additional consequences, such as increased recombination rates.

A key question is how large the effective mass is. In the following subsection we use the experimental hole-burning data to constrain the effective mass, finding that it is sufficiently large to completely rule out simple Bose-Einstein condensation of defects at the experimental temperatures.

Estimate of effective mass. We expect that the effective mass m^* of the defects is much higher than that of the free atoms. Here we bound the effective mass by considering Ahokas et al.'s measurement of the lifetime of a localized spectral hole. They found that a $w \sim 0.2\,\text{mm}$ hole persisted for $\tau_{\text{pers}} > 50$ hours. Our argument will relate macroscopic motion (which fills in the spectral hole) to microscopic motion (the defect tunneling). We assume diffusive motion, where the characteristic time between collisions is longer than a tunneling time.

There are at least three mechanisms by which the spectral hole in b atom population can heal: the excited atoms can spontaneously undergo a transition back to the b-state from the a state, spin exchange collisions can lead to spin diffusion, or b-state atoms can diffuse back into that region of space. Neglecting all but the last process gives us an upper bound on the atomic diffusion constant D of the b atoms,

$$D \lesssim \frac{w^2}{\tau_{\text{pers}}} \approx 10^{-8}\,\text{cm}^2/s. \tag{17.3}$$

Up to irrelevant numerical factors, this diffusion constant is related to the mean free path ℓ and the collision time τ by $D = \ell^2/\tau$. An upper bound for D is found by assuming that a particle undergoes a momentum changing collision every time it hops, then $\ell = d$ with d the lattice spacing and $\tau = \tau_t$ where the nearest-neighbor tunneling time is related to the effective mass by $\tau_t \sim \hbar/m^* d^2$. Since $m^* \sim \hbar\tau_t/d^2$ we constrain the effective mass of the b-atoms to be

$$m^* \sim \frac{\hbar}{D} \gtrsim 10^5 m_H. \tag{17.4}$$

where the second inequality follows from (17.3) and m_H is the bare hydrogen mass.

This effective mass is a long wavelength property of the system, and constrains the motion of atoms over long distance, effectively eliminating the possibility of global BEC. As discussed in a Sect. 17.3.2, however, on distances small compared to the mean-free path the atoms can move much more freely, and it is possible that "local" BEC/s could form, without leading to global phase coherence.

17.3.1.4 Microscopic Mechanism

Microscopic estimates of tunneling matrix elements are beyond the scope of this work. As previously discussed, the exchange reaction mechanism is expected to be the dominant pathway [6].

17.3.1.5 Experimental Predictions

Polarization temperature dependence. Perhaps the most easily testable prediction of this model is the temperature dependence of the polarization, $p = (n_a/n_b - 1)/(n_a/n_b + 1)$ with

$$\frac{n_a}{n_b} = \frac{\int d^3k\, n[\epsilon_a(\mathbf{k})/(k_B T)]}{\int d^3k\, n[\epsilon_b(\mathbf{k})/(k_B T)]} \tag{17.5}$$

with $n(x) \equiv 1/(e^x - 1)$. The energy dispersion of atoms is $\epsilon_a(\mathbf{k}) \approx \hbar^2 k^2/2m^*$, and $\epsilon_b(\mathbf{k}) \approx \hbar^2 k^2/2m^* + \Delta_{ab}$, where Δ_{ab} is the b-a energy difference. Above the BEC transition temperature T_c, the integrals yield for a homogeneous, three-dimensional gas

$$\frac{n_a}{n_b} = \frac{g_{3/2}(e^{\mu/T})}{g_{3/2}(e^{(\mu - \Delta_{ab})/T})} \tag{17.6}$$

where $g_\alpha(x) = \sum_j x^j/j^\alpha$ is the polylog function and $\mu(T)$ is chosen set the density, which yields the implicit equation for $\mu(T)$

$$\begin{aligned} &\zeta(3/2) + g_{3/2}(e^{-\Delta_{ab}/T_c}) \\ &\quad = g_{3/2}(e^{\mu(T)/T}) + g_{3/2}(e^{(\mu(T)-\Delta_{ab})/T}) \end{aligned} \tag{17.7}$$

The same expression holds in dimension d with $3/2$ replaced by $d/2$. Below $T = T_c$, one instead finds

$$\frac{n_a}{n_b} = \left(\frac{T_c}{T}\right)^{3/2} \frac{\zeta(3/2) + g_{3/2}(e^{-\Delta_{ab}/T_c})}{g_{3/2}(e^{-\Delta_{ab}/T})} - 1. \tag{17.8}$$

While the effective mass sets the density at T_c, it does not appear in this expression. The polarization depends only on Δ_{ab}/T and T/T_c. To produce the observed $n_a/n_b = 3$ at $T = 150\,\text{mK}$, one needs $T_c = 170\,\text{mK}$.

Figure 17.2 shows $p(T)$ for Boltzmann and Bose condensed (assuming $T_c = 200\,\text{mK}$) gases. Accurately measuring $p(T)$ would clearly distinguish Bose and Boltzmann statistics.

Equation 17.6 shows that Bose statistics can affect the ratio n_a/n_b even if the system is non-condensed. However, (17.6) bounds $n_a/n_b < \zeta(3/2)/g_{3/2}(e^{-\Delta_{ab}/T}) = 2.3$, which is insufficient to explain the experimentally observed $n_a/n_b = 3$.

Transport. A second signature of BEC is superflow. For example, the sample could be incorporated into a torsional oscillator, providing a measurement of a possible nonclassical moment of inertia I: below T_c, the superflow decouples from the cell, and I decreases. Mounting the sophisticated hydrogen growth and

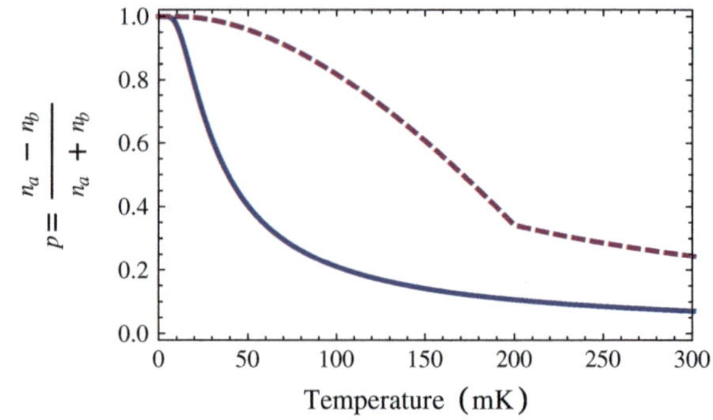

Fig. 17.2 The polarization versus temperature for the Boltzmann case (*solid line*) and the Bose-condensed case (*dashed line*), from (17.8)

measurement equipment in an oscillator would be challenging, as would the difficulty of working with such small samples.

Bimodal Cold Collision Shifts. Bose Einstein condensation also has implications for the cold collision shifts in the atomic spectra. Insofar as the interaction may be described by the s-wave scattering length the a–c line will be shifted by [15–18]

$$\delta\omega = \frac{4\pi\hbar^2}{m} g_2(0)(a_{\uparrow\downarrow} - a_{\downarrow\downarrow})\langle n\rangle \tag{17.9}$$

where $g_2(\mathbf{r}) \equiv \langle\psi^\dagger(\mathbf{r})\psi^\dagger(0)\psi(0)\psi(\mathbf{r})\rangle/\langle n\rangle^2$, $\langle n\rangle$ is the average density, and $a_{\uparrow\downarrow}, a_{\downarrow\downarrow}$ are the a–c and a–a scattering lengths, respectively. For a noninteracting BEC $g_2(0) = 1$ while for a normal gas $g_2(0) = 2$. For a partially condensed gas one in fact sees a bimodal spectrum with two peaks: one from the condensed atoms and one from the noncondensed atoms. Such bimodal spectra revealed BEC in magnetically *trapped* spin-polarized hydrogen gas [18].

Interactions between defects in the lattice are complicated, and this model of point-interactions is likely to be at best a crude approximation. In particular, atom induced lattice distortion can render the tunneling process of nearby atoms off-resonant [19], suppressing the effective g_2. The observed two-body recombination rate suggests that this suppression may be quite large. Irrespective of such strain effects, due to the symmetry of the wavefunction one would not expect to see cold collision shifts on the a–b or a–d lines [20, 21].

Thermodynamics and collective excitations. The BEC phase transition can in principle be directly observed by monitoring thermodynamic quantities such as specific heat. Due to the small number of H atoms, the signal should be quite small. Similarly, the presence of a superfluid component would lead one to expect a second-sound mode, which could be excited (for example) via localized heating of the sample.

17.3.2 *Local Bose-Einstein Condensation*

17.3.2.1 Idea

Next we pursue the idea of "local BEC", where the defects aggregate in small disconnected regions, each of which contains a condensate, but which have no relative phase coherence.

17.3.2.2 Phenomena Explainable

This model can explain the non-Boltzmann ratio n_b/n_a, and the slow transport observed in the hole burning experiments. It does not provide an explanation of the slow recombination, failure to saturate the NMR line, or fast Overhauser relaxation.

17.3.2.3 Consistency with Experiment

The arguments from Sect. 17.3.1 about the polarization go through without change. The slow recovery in the hole burning experiment is readily explained if the disconnected condensates are smaller than 0.1 mm. Furthermore, local clusters are naturally expected if there is the dramatic sort of concentrating mechanism described in Sect. 17.3.1.

17.3.2.4 Microscopic Mechanism

An attractive interaction between defects can lead to clustering. A long distance phonon mediated attraction is expected for this system [22]. Inhomogeneities in the molecular sample, or its environment could also lead to clustering. For example, it has been observed that the ortho and para molecules phase separate. Furthermore, the surface that the sample sits on creates inhomogeneous strains, which couple to defects. This can possibly cause them to accumulate. Atomic hydrogen clustering along crystalline defects—for example, dislocations and ortho-para H_2 interfaces—has been suggested to account for recombination rates and other phenomena [19, 23]. Here the motion along these defects is rapid locally, but motion over longer distances is slower, causing a higher T_c for local condensation than for global coherence among the condensates.

17.3.2.5 Experimental Predictions

The temperature dependence calculated for global BEC is unchanged for local BEC, and again a double-peaked ENDOR a-c line might emerge if microscopic interactions are favorable. Jointly observing these would provide a "smoking gun" for local BEC, but as already emphasized, strain induced interactions could readily eliminate the bimodal spectral line. We also note that at sufficiently cold temperatures the puddles phase lock giving a global BEC. The effective mass for global motion, extracted in Sect. 17.3.1 from the experimental hole burning results, constrains this transition temperature to be extremely low. The local concentration of H $\uparrow$, regardless of BEC, may be diagnosed by examining the dipolar shift of spectral lines due to H-H interactions.

17.3.3 Nuclear Spin Dependent Density-of-States

Here we outline a non-BEC scenario for the anomalous observations.

17.3.3.1 Idea

If the degeneracy of the a and b states were g_a and g_b, one would expect that $n_a/n_b = (g_a/g_b)e^{\beta\Delta_{ab}}$. Thus if a mechanism could be found to enhance (g_a/g_b), then one could explain the observed ratio of n_a/n_b. Assuming such a relative enhancement of the density of states, a strong RF field would lead to a saturated ratio $(n_a/n_b)_{\text{sat}} = g_a/g_b$.

17.3.3.2 Phenomena Explainable

Both the equilibrium n_a/n_b and the RF saturated $(n_a/n_b)_{\text{sat}}$ can be explained by this model. We also provide a scenario whereby the Overhauser relaxation is enhanced. Within this model, the low recombination rates could be a consequence of the defects being immobile.

17.3.3.3 Consistency with Experiment

At strong excitation powers the polarization $p = (n_{\text{tot},a} - n_b)/(n_{\text{tot},a} + n_b)$ saturates to

$$p_{\text{sat}} = \frac{g-1}{g+1} \tag{17.10}$$

where $g \equiv g_a/g_b$. Meanwhile, the thermal polarization is

$$p_{\text{therm}} = \frac{g\exp(\Delta_{ab}/(k_BT)) - 1}{g\exp(\Delta_{ab}/(k_BT)) + 1} \approx \frac{g-0.75}{g+0.75} \tag{17.11}$$

where the last equation holds for Ahokas et al.'s [7] experiments, where $\Delta_{ab} = 43\,\text{mK}$ and $T = 150\,\text{mK}$. If one takes $g = 2$ one finds $p_{\text{sat}} = 0.33$ and $p_{\text{therm}} = 0.45$. Ahokas et al. [7] experimentally find $p_{\text{sat}} = 0.2$ and $p_{\text{therm}} = 0.5$. One should contrast this with the naive expectation of $p_{\text{sat}} = 0.5$ and $p_{\text{therm}} = 0.14$.

17.3.3.4 Microscopic Mechanism

There are very few mechanisms whereby the molecular hydrogen matrix can change the degeneracies of the atomic hydrogen hyperfine states. The most plausible source would be to consider nuclear spin dependent interactions with o-H_2. Such interactions can be produced through spin-orbit coupling in the presence of a bias magnetic field. Hybridization of the molecular and atomic levels could in principle lead to sufficiently drastic rearrangements of the hyperfine states to affect their degeneracy. Such a strong interaction would presumably have other spectroscopic implications, such as a severe renormalization of the a–b splitting. Unfortunately, the experiments observe that the a–b splitting is changed by only 0.1% compared to its vacuum value.

If there is significant hybridization of the atomic and molecular states, then the symmetry which forbids the a–c transition would generically be broken. This would be a source of the fast Overhauser relaxation.

17.3.3.5 Experimental Predictions

One would expect that at low temperatures, g should be roughly independent of temperature. Thus the temperature dependence of the polarization in (17.11) can be compared with the experiment. For any given mechanism additional predictions are possible. As one example, in a spin-orbit mechanism hinted at above, the physics should also manifest in the populations of the o-H_2 states, which can be probed spectroscopically. Finally, under further assumptions, one may be able to predict the power dependence of the polarization saturation experiments.

17.4 Other Observations

We would like to point out three other possibly relevant observations. First, Ceperley et al. have shown that the surface of small p-H_2 clusters in a vacuum are superfluid [24]. Similarly, Cazorla et al. have shown that on small length scales 2D p-H_2 has superfluid correlations [25]. Analogous effects are predicted for defects in solid ^{4}He, including grain boundaries, dislocations, and amorphous regions [26–28]. Ahokas et al.'s crystals are of very high quality, and it is doubtful that grain boundaries and dislocations are playing an important role. On the other hand, one could imagine that the the interface of o- and p-H_2 clusters could play a similar role.

Second, it seems possible that the spectroscopic anomalies are related in some way to torsional oscillator mass decoupling observed in solid hydrogen [29] at similar temperatures. It is important to note that Clark et al. [29] ruled out *global* supersolidity in their experiments by comparing the response of the system in an open and blocked annulus. *Local* supersolidity, however, can also give rise to a small period drop, which would be present in both the open and blocked annuli.

A third observation is that magnetic ordering transitions—for example ferromagnetism—would alter the ratio of a-state to b-state population. One can eliminate ferromagnetic ordering, as it would give rise to an observable energy shift. On the other hand, perhaps a different form of magnetic order is playing a role.

17.5 Summary, Conclusions, and Consequences

We reviewed the unexplained phenomena seen in experiments of Ahokas et al. [7]. We enumerated a number of possible mechanisms which could be involved in producing the observed phenomena. In particular, we gave detailed consideration to the idea, first introduced in Ref. [7] that the non-Boltzmann ratio n_a/n_b may be due to Bose-Einstein condensation of atomic hydrogen. We conclude that global Bose-Einstein condensation is not consistent with other experimental observations.

We present several other scenarios, but find that none of them are wholly satisfactory. Even though some of the phenomena can be explained by local BEC, it fails to provide a mechanism for the unexpected saturation population $(n_a/n_b)_{\text{sat}}$ when a strong RF field is applied. We do find that all of the phenomena would be consistent with a nuclear spin dependent density of states. However, we are unable to provide a microscopic mechanism for this density of states.

Ultimately, substantial experimental work will be necessary to clarify the situation. Our arguments make a strong case that measuring the polarization's temperature dependence is a promising first step and suggests other experimental signatures—especially in transport and spectral features—that would clarify the phenomena.

During the preparation of this paper, new results came out from Ahokas et al. [30], which introduced new mysteries. In particular they observe substantial density and substrate dependence of the population ratio n_a/n_b. All of our considerations remain valid, with the additional clue that whatever the underlying mechanism is, it must involve the surface of the sample and be sensitive to density. For example, the formation of superfluid domains could be influenced by the substrate, or magnetic impurities in the substrate could interact with atomic hydrogen, leading in some way to the unexpected density of states.

Acknowledgments We would like to acknowledge discussions with David Lee, Järno Jarvinen, Sergei Vasiliev, and Cyrus Umrigar. This work was supported by NSF Grant PHY-0758104.

References

1. N. Prof'ev, Adv. Phys. **56**, 381 (2007)
2. E. Kim, M. Chan, Nature **427**, 225 (2004)
3. E. Kim, M.H.W. Chan, Science **305**, 1941 (2004)
4. A.S.C. Rittner, J.D. Reppy, Phys. Rev. Lett. **98**, 175302 (2007)
5. A.S.C. Rittner, J.D. Reppy, Phys. Rev. Lett. **97**, 165301 (2006)
6. T. Kumada, Phys. Rev. B **68**, 052301 (2003)
7. J. Ahokas, J. Järvinen, V.V. Khmelenko, D.M. Lee, S. Vasiliev, Phys. Rev. Lett. **97**, 095301 (2006)
8. S. Vasilyev, J. Järvinen, V.V. Khmelenko, D.M. Lee, AIP Conf. Proc. **850**, 81 (2006)
9. Nat. Phys. **2**, 651 (2006), doi:10.1038/nphys435
10. I.F. Silvera, Rev. Mod. Phys. **52**, 393 (1980)
11. *Quantum Tunnelling in Condensed Media*, ed. by Y. Kagan, A.J. Leggett (Elsevier Science, Amsterdam, 1992)
12. A.V. Ivliev, A.S. Iskovskikh, A.Y. Katunin, I.I. Lukashevich, V.V. Sklyarevkii, V.V. Suraev, V.V. Filippov, N.I. Filippov, V.A. Shevtsov, JETP Lett. **38**, 379 (1983)
13. N.V. Prokof'ev, G.V. Shlyapnikov, Sov. Phys. JETP **66**, 1204 (1987)
14. A.V. Ivliev, A.Y. Katunin, I.I. Lukashevich, V.V. Sklyarevskii, V.V. Suraev, V.V. Filippov, N.I. Filippov, V.A. Shevtsov, JETP Lett. **36**, 472 (1982)
15. C.J. Pethick, H. Smith, *Bose-Einstein condensation in dilute gases*. (Cambridge University Press, The Edinburgh Building, Cambridge CB2 2RU, UK, 2001)
16. M.O. Oktel, L.S. Levitov, PRL **83**, 6 (1999)
17. M.O. Oktel, T.C. Killian, D. Kleppner, L.S. Levitov, Phys. Rev. A **65**, 033617 (2002)
18. D.G. Fried, T.C. Killian, L. Willmann, D. Landhuis, S.C. Moss, D. Kleppner, T.J. Greytak, Phys. Rev. Lett. **81**, 3811 (1998)
19. T. Kumada, S. Mori, T. Nagasaka, J. Kumagai, T. Miyazaki, J. Low Temp. Phys. **122**, 265 (2001)
20. A.I. Safonov, I.I. Safonova, I.I. Lukashevich, JETP Lett. **87**, 23 (2008)
21. J. Ahokas, J. Järvinen, G.V. Shlyapnikov, S. Vasiliev, Phys. Rev. Lett. **101**, 263003 (2008)
22. G.D. Mahan, (2010)
23. V. Shevtsov, A. Frolov, I. Lukashevich, E. Ylinen, P. Malmi, M. Punkkinen, J. Low Temp. Phys. **95**, 815 (1994)
24. S.A. Khairallah, M.B. Sevryuk, D.M. Ceperley, J.P. Toennies, arxiv 0612161 (2006)
25. C. Cazorla, J. Boronat, arXiv:0807.0307v1 (2008)
26. L. Pollet, M. Boninsegni, A.B. Kuklov, N.V. Prokof'ev, B.V. Svistunov, M. Troyer, Phys. Rev. Lett. **98**, 135301 (2007)
27. M. Boninsegni, A.B. Kuklov, L. Pollet, N.V. Prokof'ev, B.V. Svistunov, M. Troyer, Phys. Rev. Lett. **99**, 035301 (2007)
28. Ş. G. Söyler, B. Capogrosso-Sansone, N.V. Prokof'ev, B.V. Svistunov, Phy. Rev. A (Atomic, Molecular, and Optical Physics) **76**, 043628 (2007)
29. A.C. Clark, X. Lin, M.H.W. Chan, Phys. Rev. Lett. **97**, 245301 (2006)
30. J. Ahokas, O. Vainio, S. Novotny, J. Järvinen, V.V. Khmelenko, D.M. Lee, S. Vasiliev, Phys. Rev. B **81**, 104516 (2010)

Chapter 18
Helium and Hydrogen (super?)Solids

18.1 Background

Supersolidity is a counterintuitive phenomena where a solid—a substance resistant to shear—also allows superflow—dissipationless flow to pressure gradients. The discovery of anomalous rotational properties in ^{4}He was consistent with supersolidity [1, 2], but further investigation indicated that the simple pictures that had been developed were insufficient to account for the data.

The work in this chapter was largely carried out in 2005–2006, when a broad range of new experimental results complemented the pioneering torsional oscillator studies, and we tried to account for many of the seemingly contradictory observations. It outlines a scenario that could be consistent with all of those observations—but makes no claim of any substantial likelihood of correctness. At best, I hoped to stimulate a simpler, more complete account of the phenomena.

In the time since this period, many additional experiments have been performed and some of the experimental "facts" have been called into question. Thus, although the scenario was constructed for consistency with the data at the time, the experimental state of the art makes it unclear whether it is consistent with all of the most up to date observations.

Regardless, our scenario is interesting as a possible manifestation of supersolidity in disordered systems: in particular, it connects microscopic calculations (see [3] and references therein) with the macroscopic behavior probed in experiments, and due to these ideas and this connection it may be useful in other contexts.

18.2 Chapter Overview

I develop a phenomenology of a particular model of disordered supersolidity. It is related to proposals that superflow occurs along defects—dislocations, grain

K. R. A. Hazzard, *Quantum Phase Transitions in Cold Atoms and Low Temperature Solids*, Springer Theses, DOI: 10.1007/978-1-4419-8179-0_18,

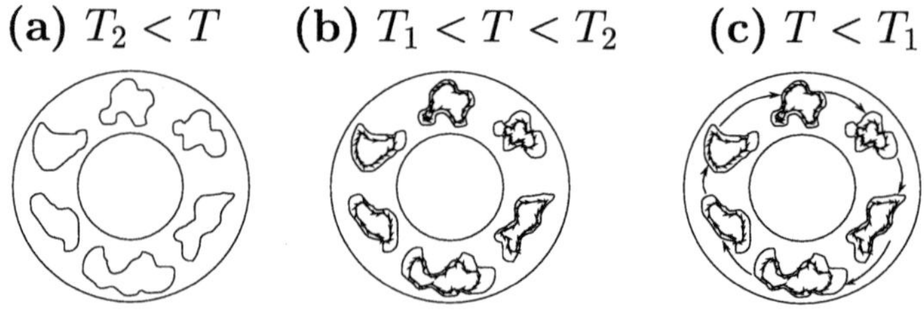

Fig. 18.1 Our model's basic idea. **a** The solid at $T > T_2$ before any superflow develops. There are two distinct types of regions: those that support superflow at low temperatures (SF) and those that will not (NSF). It is likely that T_2 is larger than the solidification temperature so that the domains are *always locally superfluid* in the solid. **b** The state of the system after cooling through one phase transition to a temperature $T_1 < T < T_2$: within each of the SF regions, a superflow develops. However, no global phase coherence is present. **c** The state of the system after cooling to $T < T_1$: global phase coherence has developed, and superflow occurs throughout the system

boundaries, and amorphous regions—but develops a picture of the macroscopic, experimental properties of the system starting from these microscopic considerations. The essence of the idea is that superflow may exist in large disordered regions, and that the supersolid feature observed in torsional oscillator experiments corresponds to the point where these become globally phase coherent, even though the superflow. *within* the domains is already present at higher temperatures. The experimental consequences of this are fleshed out and found to be consistent with the present state of experiments.

18.3 Introduction

A number of conundrums surround supersolid ^{4}He [3]. As some notable examples: one finds a drastic *increase* in T_c with ^{3}He doping [1, 2, 4], a mysteriously similar non-classical rotational susceptibility in H_2 which shows none of the other signals of superflow [5], frequency dependence of the transition temperature [6], and anomalously small thermodynamic features (e.g., heat capacity peak) [7] one expects to be associated with superfluidity. We describe a model of defects that are superfluid even at much higher temperatures than the supersolid transition, but become globally coherent only at the supersolid transition temperature T_c.

This theory allows an understanding of all these experimental mysteries.

Other unexplained experimental features exist: the geometry dependence of the supersolid fraction [8–10], an anomalously low critical velocity, absence of dc flow [11] driven by external pressure gradients, low-temperature specific heat and pressure data [7], and anomalous shear modulus stiffening at the supersolid transition [11]. While our theory is not necessarily linked with each of these, none of these results are immediately inconsistent with our ideas.

Figure 18.1 illustrates the basic idea of our model graphically. In words, it is that there exist two phase transitions, at T_1 and T_2 with $T_1 < T_2$. Some domains of the solid turn into independent supersolid domains (domains that support superflow of some kind within the domain) as the temperature is lowered below T_2. Then, upon further cooling below T_1, phase coherence develops between the independent superfluid regions allowing a global superflow. *It is likely that the local superfluidity occurs at all temperatures in the solid, in which case there will be no observable* T_2*"transition"*.

Given the observed robustness of the transition—its existence is relatively independent of sample and geometry—it is necessary that the links connecting the supersolid regions are somehow robust, most likely independent defects, and not, say, to just tunneling through the solid.

Before examining its consequences, it is worth seeing what this theory is and what it is not. It is not an *ab-initio* theory predicting the mechanism of supersolidity. Rather, it takes as given the point of view advocated by many that that there are regions of the crystal which can maintain superflow. From this, it derives the macroscopic observable quantities, in the process clarifying several puzzling experiments. We now turn to these mysteries individually.

18.4 Torsional Oscillator NCRI

Torsional oscillator experiments are a classical test of superflow. Basically, a cell of some geometry filled with solid helium is oscillated at a frequency ω about its center—an annulus, as assumed in the illustration in Fig. 18.1, a cylinder, and an extruded rectangular cell have all been used in the study of supersolid helium and hydrogen. As $\omega \to 0$, elementary hydrodynamics shows that a normal fluid will follow the walls of the container,as will a solid. However, for a system supporting superflow, as $\omega \to 0$, a finite fraction of the sample does not oscillate with the walls and the effective moment of inertia is less than that of the system without superflow.

In torsional oscillator experiments, the system is driven on-resonance with a lock-in amplifier. The period is $T \propto \sqrt{I}$. So as the resonance frequency changes, one can infer the change in I and hence the superflow fraction. Furthermore, the amplitude at resonance is proportional to the quality factor Q. Thus the loss via the Q factor and the moment of inertia via T may be measured.

18.5 Two Supersolid Features

At each of T_1 and T_2 we expect to see a period drop going from high to low temperature, however *we suspect that* T_2 *may be above the solidification temperature, and thus non-visible*. Even if T_2 were below the solidification temperature, the feature would be observable only in very large annuli, as small disconnected regions give only a small contribution to the superflow. The square cell Cornell experiments have indeed seen two features [8],[1] the higher temperature of which has yet to be explained, although this is probably an artifact. For reference, these occur at roughly 150 mK and 1 K. There are hints that Goodkind's ultrasound attenuation and sound speed data also sees a high-temperature transition, as well, though the data is not easy to interpret [12].

For small domains of characteristic size r, the period drop associated with the high-temperature feature gives an apparent supersolid fraction going as $(r/R)^2$ where R is the radius of the cell. (This calculation appears elsewhere.) In annuli, the annular width limits the maximum r and hence limits the apparent supersolid fraction upon going through T_2 to be very small.

18.6 Blocked Annulus Torsional Oscillators

Blocking an annular cell prevents global superflow. Some superflow pattern will be established in the cell that leads to a moment of inertia reduction, but it will be much less than the naive "fraction of superflow equals fraction of mass decoupling" that applies in the unblocked case. Typical cases for current annular geometries involve a period drop for the blocked annulus that is roughly $\sim 1/100$ of the unblocked geometry, though the blocked apparent supersolid fraction can be as large as ~ 0.5 of the unblocked fraction for a very large annular width .

When the annulus is blocked, the isolated superfluid regions, whose condensation is associated with T_2, are going to be hardly affected. On the other hand, the superflow *around* the annulus associated with the tunneling between the independent regions occurring below T_1, will be unable to establish its usual flow pattern.

Consequently, if prior to blocking there is a visible transition at T_2, then upon blocking, the period drop for the transition at T_1 should decrease roughly in accordance with a blocked uniform superflow, while the period drop at T_2 should change negligibly. Experimentally, such a drop is indeed observed in ^{4}He for the T_1 transition. In Sect. 18.8 we argue that this phenomenon explains Chan's H_2 solid results [5] in a natural way.

18.7 Dissipation Peaks

All experiments observe a dissipation peak accompanying the period drop. Such a dissipation maximum is not expected in the context of the usual superfluid transition. However, such a feature is a standard characteristic of the two-dimensional

---

[1] Sophie Rittner and John Reppy, private communication

Kosterlitz–Thouless transition occurring in superfluid films.This suggests that—in some respects—the flow could be two-dimensional.

Several recent works, both theoretical and experimental, have shown that grain boundaries in ^{4}He support superflow, giving a natural candidate for two-dimensional flow. The transition in the grain boundaries then may account for the dissipation feature. Moreover, two-dimensional systems have a relatively smeared out period drop; this is also present in the supersolid experiments, but this may arise also due to a disorder-induced smearing of transition temperatures.

Nevertheless, a mechanism with solely two-dimensional flow is hard to reconcile with the large supersolid fractions recently observed by Rittner and Reppy (RR) [10]. In order to understand this better, it is worth examining the experimental correlation between dissipation and apparent supersolid fraction. Although RR were able to increase the supersolid fraction by orders-of-magnitude through quenching and restricting the sample geometry, the dissipation does not increase commensurately. This suggests that in samples with more than a few percent supersolid fractions, a substantial portion comes from non-two-dimensional objects, while two-dimensional objects account for a respectively larger portion at smaller supersolid fractions. Although the dissipation feature is not required by our theory, it gives one ostensible interpretation.

18.8 H_2 Experiments

In molecular H_2 solid experiments [5], a feature reminiscent of that in ^{4}He is observed at roughly 120 mK. A *natural* way of interpreting the experiment is to identify it as a supersolid transition of some kind. However, there are at least five mysterious aspects of this experiment:

- The signal does not disappear after blocking the annulus.
- The temperature of the measured ^{4}He T_1 transition coincides with that of H_2, but as indicated by the de Boer and Lindemann parameters, ^{4}He is "more quantum" and thus presumably should have a higher transition temperature.
- There is no critical velocity up to 500 μm/s in H_2.
- The feature is not present when HD is substituted for H_2.
- There is no dissipation maximum in the H_2 data, while there is for the T_1 transition in ^{4}He.

The present analysis can account for all of these phenomena, relate parameters to the ^{4}He experiments, and make "easily" testable predictions.

Reference [5] attempts to explain some of these mysteries via clustering of ortho- and para-hydrogen in specific locations of the cell. However, this theory is somewhat unnatural in providing no reason that the effect and transition temperature are so similar to the ^{4}He transition, nor does it provide a reason the cluster would occur in the requisite extreme manner.

There is also a serious difficulty with the clustering explanation proposed in the preprint in that *it is impossible to account for the magnitude of the effect* by clustering: even the most extreme clustering falls short. Our scenario does not suffer these defects.

Let us examine the first of the mysteries, that *the superflow fails to disappear upon blocking the annulus. Such a feature is expected within a theory of disordered domains*, when one is above the global coherence transition. The observed feature, then, would be associated with the "T_2" feature.

Next, *the experimental feature disappears if* HD *is substituted for* H_2, *consistent with our theory, since* HD *is a fermion*, consistent with our picture, but not easily with other pictures.

Finally we will consider the transition temperatures. The Lindemann ratio and the de Boer parameter are, in this context, ways of characterizing the "quantumness" of a solid. The former is defined as the ratio of the zero-point root-mean-displacement of an atom of the solid to the interatomic spacing. "Classical solids" have a more or less universal ratio of around 10% for melting. H_2 and ^{4}He, the two atomic quantum solids, have Lindemann ratios of 18 and 26, respectively, in their stable ground states. Importantly, ^{4}He has a larger Lindemann ratio and hence is "more quantum"—we would thus expect the supersolid transition at a *higher* temperature than H_2.[2] The de Boer parameter also supports this reasoning: it is defined as $\Lambda \overset{\text{def.}}{\equiv} \frac{h}{\sigma\sqrt{m\epsilon}}$, with σ the equilibrium spacing of the fit Lennard–Jones potential and ϵ the the dissociation energy. For intuition, note that the de Boer parameter is essentially capturing the same thing as the Lindemann ratio: it is the ratio of the length scale associated with the dissociation energy—$h/\sqrt{m\epsilon}$-to the interatomic spacing σ; they are simply different ways of viewing the same physics of zero-point fluctuations. For H_2 and ^{4}He, the de Boer parameters are 1.7 and 2.6, respectively. Again, ^{4}He is "more quantum," with greater zero point motion, and hence should show superflow at a higher temperature.

In short, this suggests that H_2 should become a supersolid at a lower temperature than ^{4}He, in stark contrast to the naive interpretation of the observations, where they occur at roughly the same temperature. However, in light of the arguments earlier in this section, this becomes less puzzling: we should not compare the feature in H_2 with the usual ^{4}He supersolid feature, but rather with the a "T_2" feature. Assuming $T_{2,{}^4\mathrm{He}} \sim 1$K, very roughly on the order of the melting temperature, we see that the transition temperature observed in H_2 is indeed significantly lower than the corresponding transition in ^{4}He.

The Lindemann ratios of ^{4}He and H_2 provide a simple estimate of the ratio of their respective transition temperatures. The zero point motion of the atoms is the standard deviation of position for that atom; taking a simple cosine as a model for the lattice potential for simplicity allows one to calculate the Wannier functions as

---

[2] Assuming that the mechanism—defects carrying superflow, etc.—in the H_2 is analogous to that in ^{4}He.

a function of lattice depth and spacing; then given the experimental values of lattice spacing and Lindemann ratio we can determine the lattice depth, giving a complete characterization of the potential from which we calculate the hopping energy of nearest neighbors. Note that the hopping energy/rate must be on the order of the transition temperature, since it is the only energy scale in the (non-interacting) system. Then this estimate produces a ratio of transition temperature

$$\frac{T_{^4\mathrm{He}}}{T_{\mathrm{H}_2}} \sim 2.1$$

in rough agreement with the observed factor of 4 or 5. We would not consider taking such a simplistic, toy estimate seriously if it were not for the fact that it gives results in qualitative agreement with experiment.

Next, *the theory accounts for the apparent lack of a critical velocity in* H_2. In an annulus of radius R with oscillation frequency and amplitude of angle oscillation Θ_0, the angle of the system is $\Theta(t) = \Theta_0 \sin(\omega t)$, while the rim velocity as a function of time is $v_{\mathrm{rim}}(t) = \Theta_0 R\omega \cos(\omega t)$. We can assume this velocity to be uniform throughout the sample for a thin annulus. The maximal velocity is then $v_{\mathrm{m}} = \Theta_0 R\omega$. It is a standard feature of the theory of superflow that the superflow disappears when the superflow moves relative to the background wall and normal fluid with a velocity greater than a critical velocity v_{c}. In the usual case of uniform superflow in the annulus, this corresponds to

$$v_m = v_c. \tag{18.1}$$

However, in the case of present interest, where individual domains support superflows locally, but without global coherence, the flow velocity relative to the background is clearly not v_m. To calculate the velocity relative to the background (cell and normal solid), it is easiest to work in the rotating reference frame where the background does not move. The hydrodynamic solution for the superflow of a circular domain in the lab frame, being irrotational, yields a flow that translates in space, but has no rotation about the center of mass. So in the rotating frame the center of mass is stationary, and there is rotation of the flow about the center of mass. In the rotating frame, this rotation goes through an *angle* that is the same in magnitude and opposite in orientation to the angle traversed in the lab frame by the domain's center of mass. If the circular domain has radius r, then, the angle in the domain through which the flow occurs is $\Theta_{\mathrm{dom}} = -\Theta_0 \sin(\omega t)$ and maximal velocity in the domain relative to the background is thus $v_{\mathrm{dom}} = \Theta_0 r\omega$. The critical velocity occurs then at $v_{\mathrm{dom}} = v_{\mathrm{c}}$. Comparing this with (18.1), we see that *the critical oscillation velocity in the disconnected case is enhanced by a factor of* R/r *to the connected case's global flow.*

This can be re-phrased in an illuminating way: in the connected case the breakdown of superflow happens at

$$v_m = v_c; \tag{18.2}$$

in the disconnected case it occurs as

$$v_m = (R/r)v_c. \tag{18.3}$$

Define a frequency v_m/R and a critical frequency $\Omega_C \overset{\text{def.}}{\equiv} v_C/r$: then we see that *the critical breakdown of superflow in the disconnected case occurs at a critical* frequencyΩ_c independent of the system geometry, rather than the usual critical velocity.

Applied to [5]'s H_2 experiments, we find that the apparent critical velocity will be enhanced by a factor $R/r > 5.6$ relative to the true critical velocity (this follows since the domain radius must be at most half of the annular width and from using the geometry of those experiemnts). Additionally considering that the 10–100 μm/s critical velocity in ^{4}He is abnormally low, *one does not expect a breakdown offlow in* H_2 *up to* 500 μm/s *within our scenario.*

It is also worth estimating the true supersolid fraction of the domains needed in order to generate the apparent global supersolid fraction of $\lesssim 0.1\%$. It is clear that the ratio of the apparent supersolid fraction in this picture to the "true" supersolid fraction (i.e., the supersolid fraction if things were globally connected) is the same as the ratio of the high-temperature feature's period drop to the total period drop (calculated elsewhere) This is

$$\frac{\rho_{\text{app}}}{\rho_{\text{true}}} = \frac{1}{1 + 2\left(\frac{R}{r}\right)^2}.$$

Because the circular domains can have radii no larger than the annular width divided by two, we then know that the this ratio can be as large as 0.015. In order to obtain the observed apparent fraction of $\sim 0.1\%$ from domains of this size, one needs a not unreasonable domain supersolid fraction of $\sim 7\%$. Indeed, larger supersolid fractions have been observed in ^{4}He by RR. Nevertheless, the NCRI decreases rapidly with decreasing domain size, so if our mechanism is correct, the domains would need to be very near their maximum size and the "true" supersolid fraction to be at least a few percent.

Actually, there is a caveat to the limit argued in the last couple paragraphs and the apparent supersolid fraction generated by this mechanism can be even higher than estimates. Most importantly, the cell is constructed with radial channels and a central fill line; the majority of the feature's magnitude can be accounted for just by assuming there is a superflow in the radial channels even if there is none in the annulus or fill line—clearly such a flow would not disappear upon blocking the radius.

18.9 Frequency Dependence

Typical theories involving disorder and glassy dynamics predict a frequency dependence of the superflow fraction below the transition temperature. In contrast, the Rutgers experiments—in oscillators with two simultaneous resonant frequencies—have showed a superflow fraction independent of frequency [6]. However, they displayed at least two intriguing features: (1) at finite T a frequency-dependent critical velocity (in contrast to usual superflow) that becomes frequency independent at low T, and (2) an intriguing shift of T_c with frequency.

Both anomalies are consistent with our model with dislocation links—or more generally, any fluctuating links. At zero temperature, the links are frozen and one probes the intrinsic dynamics of the superflow. One then sees a frequency-independent critical *velocity* as expected. At higher temperatures however, the links become more easily displaced, and the finite frequency probes the dynamics of the links as well as the superflow, so one no longer observes a strictly frequency-independent velocity dependence of the superflow fraction.

Similarly analyzing the frequency dependent T_c in our model, the links dynamics give a frequency dependence to the amount of time spent near domains, and hence to the effective tunnelling rate, and consequently temperature.

It should be possible to make precise predictions within our theory regarding the magnitude of such shifts, once one commits to a specific model of the links between domains. This would provide a very direct experimentally testable prediction.

18.10 Thermodynamics: Specific Heat and Pressure

First, let us start by reviewing the experimental status of thermodynamic measurements in ^{4}He [7]. *There are two sides to the mysteries: on the one hand, there is no heat capacity feature at the supersolid transition, as there would be in a naive theory of Bose-Einstein condensation; on the other hand, there is some rather peculiar behavior going on at temperatures well below the supersolid transition,* with the pressure varying as $P \sim T^2$ and the specific heat either going as $C \sim T$ or having a feature interpretable in terms of an "excess specific heat." Also, there is an absence of any pressure signature of the transition along the melting curve.

Let us start with the second mystery. The relations $P \sim T^2$ and $C \sim T$ are a generic feature of systems with a non-zero density of states at zero energy. *Systems with disorder generally will have a non-zero density of states at zero energy, so both the specific heat and pressure temperature dependencies are expected from our theory*. This is consistent with, but hardly unique to, our theory—virtually any theory involving disorder will naturally be consistent with these results.

The more interesting observation is the absence of a specific heat peak. Consider an ordinary superfluid; upon cooling through the superfluid transition, a fraction of the liquid goes from carrying the usual entropy of a liquid to carrying no entropy, since the superfluid component exists in a single quantum state. This generates a peak feature at the superfluid transition temperature as defined via the mass decoupling/non-classical rotational inertia. However, in our phenomenology, at the supersolid transition each domain has a fully developed superflow, with only global coherence absent. More precisely, for usual superfluidity, above the transition *each particle* has a few degrees of freedom associated with it, and hence the entropy drop upon fully developing superfluidity is on the order of k_B times the number of particles. In contrast, in our scenario, above the transition, the entropy is determined by noting that *each domain* will have only a few states associated with it (corresponding to its indeterminate phase) that change upon going through the global coherence transition. Hence the entropy drop from this point to fully developed globally coherent superflow is on the order of k_B times the number of *domains*. *Since the number of domains is drastically less than the number of particles, there any specific heat feature should be very small, on the order of k_B per domain.* In fact, a peak with $\sim k_B$ per ^{3}He impurity has been observed.

18.11 ^{3}He Doping

Yet another very puzzling feature of the ^{4}He experiments is that there supersolid feature occurs at *higher* temperatures as one *increases* the (fermionic) ^{3}He concentration from the isotopically pure sample. The supersolid fraction increases as well for small ^{3}He concentrations, before decreasing starting at concentrations around 100 parts per billion. Contrast this to a superfluid where such small amounts of ^{3}He doping have little effect, and where the first noticable effects are a *decrease* in superfluid fraction and transition temperature.

We can very tentatively suggest a semi-quantitative explanation of these facts within our theory. Let us adopt a more specific picture than we have previously: consider that the ^{3}He serves as a nucleation site for the domains; this is physically plausible since the ^{3}He is distinct from the ^{4}He and has a lower mass, and hence greater zero-point fluctuations, creating pressure gradients near the ^{3}He and presumably a nucleation site for disorder. We have been assuming that there is a backbone connecting these domains (perhaps a grain boundary network).

We will examine two, essentially distinct, mechanisms for the physics of the links: (1) a model of frozen links where T_c is associated with the tunneling rate through the links and (2) a model of dynamic links where T_c is associated with the binding of links to domains. Each give adequate descriptions of the data, although the latter agrees remarkably well.

18.11.1 Frozen Link Model

In a model with domains connected by a backbone of links, the tunneling rate between the domains is

$$\Gamma = t_c^2 G \tag{18.4}$$

where t_c is a matrix element for tunneling from a domain onto the backbone and where G is the matrix element for propagating along the backbone from the location of one domain to another. For a two-dimensional condensed system the static Green's function to propagate from a point $\mathbf{x}_1$ to $\mathbf{x}_2$ is proportional to $1/\sqrt{d}$ with the d the distance between domains

$$d \stackrel{\text{def.}}{\equiv} \sqrt{|\mathbf{x}_1 - \mathbf{x}_2|}, \tag{18.5}$$

basically from probability conservation. (This happens even in a non-condensed system, so long as the particles propagate ballistically; i.e., the mean free path is longer than the distance between domains). Then the tunneling rate between two domains is

$$\Gamma \propto 1/\sqrt{d}. \tag{18.6}$$

For the simple model with amorphous regions nucleating near ^{3}He, we can obtain d from the domain ^{3}He density. If the ^{3}He is uniformly distributed, then $d \propto n_{^d\text{He}}^{1/3}$. On the other hand, it is known that ^{3}He likes to cluster around surfaces or possibly, due to the elastic strain, grain boundaries as well. In this case, $d \propto n_{^d\text{He}}^{1/3}$. Then

$$\Gamma \propto n_{^3\text{He}}^x \tag{18.7}$$

with $n_{^3\text{He}}$ the ^{3}He density and $x = 1/6$ for uniform distribution of ^{3}He and $x = 1/4$ for uniform distribution along a two-dimensional object. The prediction is shown in Fig. 18.2; we see that the prediction is quite good for both power laws, but especially for the $x = 1/4$. (Notice that the graph is a log–log plot spanning five decades of concentration!). This is an interesting conclusion as fractional power laws for temperature versus concentration are somewhat non-trivial.

18.11.2 Dynamic Link Model

In this scenario, we suggest the supersolid transition T_c occurs when the links (most likely dislocations) bind to the domains. One such cause could be dislocation links binding to amorphous regions, in turn nucleated on ^{3}He sites. Such physics is essentially equivalent to Ref. [4]'s model for dislocation-^{3}He interaction turned on its head: now there are more or less static domains associated with the ^{3}He to which vibrating dislocations are pinned. As Kim et al. argue, the transition temperature for this pinning is

$$T_c = -2E_B \left(\ln \left[\frac{x_3^2 L_{\text{IP}}^3 E_B}{4\mu b^6} \right] \right)^{-1} \tag{18.8}$$

where E_B is the binding energy of the dislocation to the domain, L_{IP} is the distance between intrinsic pinning nodes of the dislocation network, μ is the ^{4}He bulk shear modulus, and b is the magnitude of the dislocation's Burger's vector. Figure 18.2

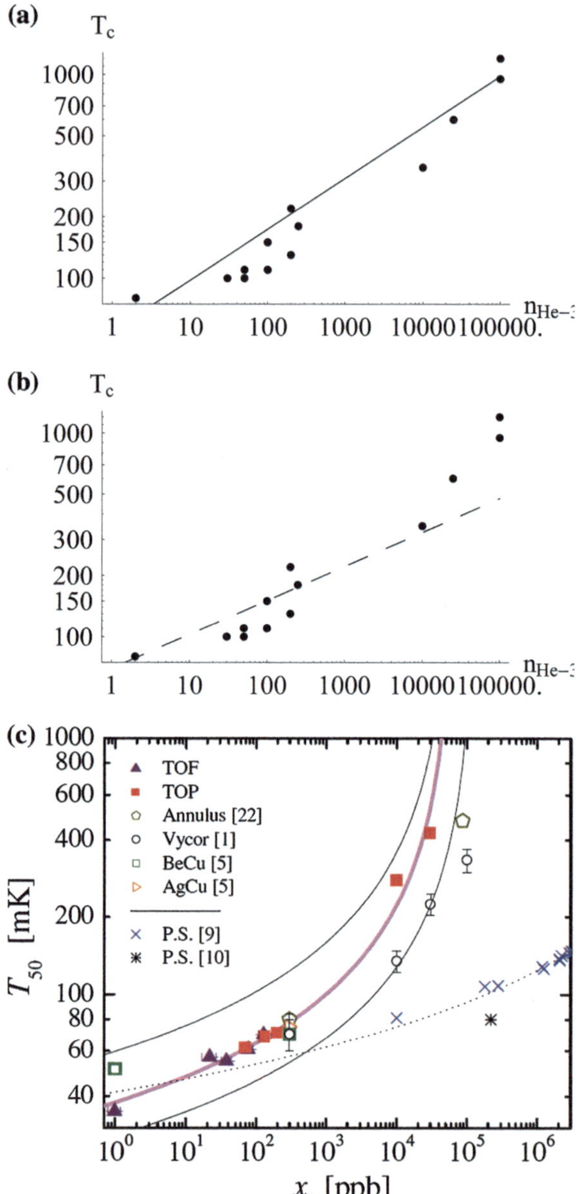

Fig. 18.2 The critical temperature (in mK) versus ^{3}He concentration (in parts per billion) on a log–log plot. *Solid line*. $x = 1/4$ static links prediction (see text); *Dashed line* $x = 1/6$ static links prediction. (Points are copied by hand out of Moses's proposal.) The color figure is from Kim et al. [4] and compares various experimental T_c's with (18.8), our prediction for T_c (*solid pink line*)

shows that this fit, taken from Ref. [4] is in good agreement with the data. The fitting parameters also yielded physically reasonable values.

Note that while this formula was given in Kim et al. they offered no explanation of how it leads to the observed "supersolid" features. Here we note that it may signify the temperature at which local *superflow becomes global, due to the binding of links that support superflow between domains.*

Regardless of the mechanism considered, *the maximum temperature which the supersolidity has reached is roughly the theoretically predicted values for defects*, generally between 0.5 and 1.5 K for grain boundaries, glassy regions/supercooled liquid, etc. *This is what is expected if increasing* ^{3}He *brings the domains closer together.*

18.12 Anomalous Critical Velocity

The theoretical critical velocity in known supersolid region candidates (grain boundaries, glassy regions, etc.) is on the order of 1 m/s. However, the experimentally observed critical velocity as measured in the torsional oscillator experiments is between 10 and 100 μm/s.

One idea consistent with our theory is that this reduction is simply due to a variety of orientations of line or plane defects that carry supercurrent. Then for a velocity $v_{\rm osc}$ in the oscillating frame, defects which are oriented an angle θ away from the direction of oscillation will have velocity along them of $v_{\|} = v_{\rm osc}/\cos\theta$, so that the superflow will breakdown at a critical $v_{\rm osc}/\cos\theta$. Since $v_{\|} > v_{\rm osc}$, the apparent critical velocity deduced from the critical $v_{\rm osc}$ will be much lower than the critical velocity for flow along the defect.

18.13 DC Flow

Here we try to understand some of the distinctions between different classes of dc flow experiments within our theory; in particular why do experiments observe no dc superflow?

The basic idea comes from UMass in a vivid slogan: one cannot squeeze a fluid from a stone. Basically, any externally applied pressure that would drive flow need not be transferred to the fluid carried by the defects, if the defects are strong enough to support a pressure gradient across them, to "protect" the superflow.

If this is the case, the addition of a reservoir of superfluid to the left will allow mass flow, but there is no such reservoir in the Beamish experiments. An experiment with a reservoir has been performed by Hallock's UMass experimental group, giving apparent flow [13, 14]. Nevertheless, some aspects of this flow remain puzzling.

18.14 Shear

Day and Beamish have observed a characteristic stiffening of the solid at low temperatures, with an appropriately rescaled shear modulus displaying an essentially identical feature as the torsional oscillator period drop [11]. A proposed interpretation is that this is a dislocation pinning phenomena. Indeed, this hints at the picture presented above, where the ^{3}He concentration dependence of T_c comes from binding of dislocations to ^{3}He or amorphous regions. In this case, one expects the binding, etc. to continue even in the absence of superflow (e.g., for ^{3}He). This is observed in experiment is consistent with our theory.

18.15 Implications for Mechanism

We have interspersed two arguments: first, those regarding our general scenario and secondly, postulating more particular mechanisms. In the process, a consistent, specific mechanism emerged: local ^{3}He-induced 3D defects supporting superflow—perhaps superglass—that anneal away, connected by a lower dimensional—perhaps dislocation or grain boundary—backbone in ^{4}He. In H_2 we argued that the domains never connect (become globally coherent) at the relevant temperatures.

18.16 Future Directions

To re-iterate our general scenario, on both sides of the apparent supersolid transition in ^{4}He, each domain is supporting a finite superflow. At the transition, enough links support superflow that a globally connected cluster forms. Moreover, it is most plausible that all of these links begin to connect at roughly the same

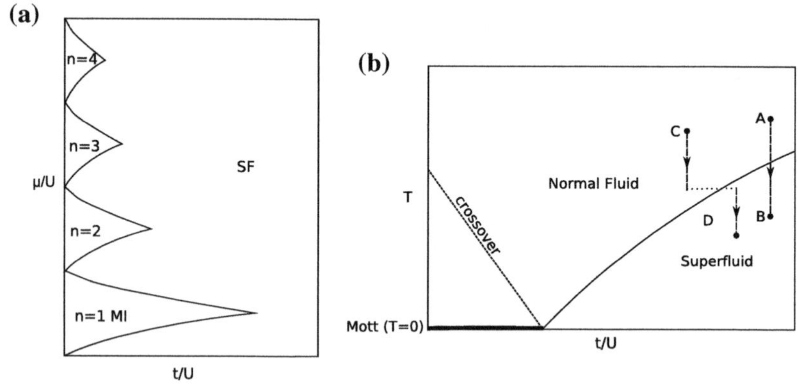

Fig. 18.3 A schematic review of the Bose–Hubbard model: The phase diagrams for the Hubbard or coupled Josephson junction model. Here t is the tunneling rate, μ is the chemical potential, U is a measure of the interaction strength between particles on the same site, and T is the temperature. The "$n = x$" states are the x-particle per lattice site Mott insulating states. (**a**) Zero-temperature phase diagram, consisting of Mott insulating lobes with integer fillings and a superfluid phase; presumably the supersolid phase is far from the Mott lobes. (**b**) The finite temperature phase diagram, with a normal fluid and a superfluid phase. While $A \rightarrow B$ would be a cooling path for a system directly described by the Hubbard model, our theory associate a more indirect path in t/U with the experimental cooling, as illustrated by the path $C \rightarrow D$

temperature, given the robustness of the transition temperature to sample-to-sample variation, annealing, and sample geometry.

Such a system is, at least qualitatively, described by a Bose–Hubbard model or, equivalently, coupled Josephsons junctions on an irregular graph where the vertices represent superfluid domains and the edges represent tunnel-coupling between the domains. Note that this is a Bose–Hubbard model with many particles, perhaps millions, per site, but this does not change the analysis qualitatively. For reference, Fig. 18.3 sketches the phase diagrams at both zero and finite temperature. Figure 18.3b illustrates discusses possible effects of temperature in this model.

From such a model one might be able to predict off-resonant frequency dependence of the observations, and design novel experiments to more precisely test the ideas presented in these notes, for example in the multi-frequency oscillator experiments.

18.17 Conclusions

Despite it's phenomenological nature, we see our picture clears up a number of experimental mysteries and, if verified, it offers a method of examining microscopic parameters. Through this last feature, it may shed light on the *microscopic* mechanism of superflow, and a plausible suggestion for the mechanism has been given here in terms of amorphous regions connected by a grain boundary or surface "backbone".

KH acknowledges stimulating discussions with S. Basu, J. Beamish, K. Shirahama, N. Prokof'ev, and J. Sethna. We are especially grateful for discussions and receiving (at the time) unpublished results from A. Rittner, J. Reppy, and M. Chan.

References

1. E. Kim, M. Chan, Nature **427**, 225 (2004)
2. E. Kim, M.H.W. Chan, Science **305**, 1941 (2004)
3. N. Prokof'ev, Adv. Phys. **56**, 381 (2007)
4. E. Kim, J.S. Xia, J.T. West, X. Lin, A.C. Clark, M.H.W. Chan, Phys. Rev. Lett. **100**, 065301 (2008)
5. A.C. Clark, X. Lin, M.H.W. Chan, Phys. Rev. Lett. **97**, 245301 (2006)
6. Y. Aoki, J.C. Graves, H. Kojima, Phys. Rev. Lett. **99**, 015301 (2007)
7. X. Lin, A.C. Clark, Z.G. Cheng, M.H.W. Chan, Phys. Rev. Lett. **102**, 125302 (2009)
8. A.S.C. Rittner, J.D. Reppy, Phys. Rev. Lett. **97**, 165301 (2006)
9. A.S.C. Rittner, J.D. Reppy, Phys. Rev. Lett. **98**, 175302 (2007)
10. A.-S. Rittner, J. Reppy, Phys. Rev. Lett. **101**, 155301 (2008)
11. J. Day, J. Beamish, Nature **450**, 853 (2007)
12. J.M. Goodkind, Phys. Rev. Lett. **89**, 095301 (2002)
13. M.W. Ray, R.B. Hallock, arxiv:1007.2156
14. M.W. Ray, R.B. Hallock, arxiv:1005.1022

Appendix A
Relating Scattering Amplitudes and T-Matrix

I will derive how the scattering amplitude is related to the T-matrix. First, I will derive the equation satisfied by the scattering amplitude, and show that this is the same as the Lippmann-Schwinger equation satisfied by the T-matrix. This allows me to relate the scattering length a to the T matrix at zero momentum.

To see that the scattering amplitude and the T-matrix satisfy this equation, we use the definition of the scattering amplitude along with the fact that ψ satisfies the Schrödinger equation. Since in momentum space ψ satisfies

$$\frac{k^2}{m}\psi(\mathbf{k}) + \frac{1}{\Omega}\sum_{\mathbf{k}'} V(\mathbf{k}-\mathbf{k}')\psi(\mathbf{k}') = E\psi(\mathbf{k}) \tag{A.1}$$

with Ω the volume of space. For our potential,

$$\frac{k^2}{m}\psi(\mathbf{k}) + \frac{4\pi a}{m\Omega}\sum_{\mathbf{k}'} \psi(\mathbf{k}') = E\psi(\mathbf{k}). \tag{A.2}$$

The scattering wavefunction for incoming wavevector $\mathbf{k}$ is defined so that

$$\psi_{\mathbf{k}}(\mathbf{k}') = \psi_{0,\mathbf{k}}(\mathbf{k}') + \psi_{\text{scat},\mathbf{k}}(\mathbf{k}') \tag{A.3}$$

where $\psi_{0,\mathbf{k}}$ is the wavevector $\mathbf{k}$ solution to the Schrödinger equation for $a = 0$. Thus,

$$\begin{aligned} &\frac{k'2}{m}\left[\psi_{0,\mathbf{k}}(\mathbf{k}') + \psi_{\mathbf{k},\text{scat}}(\mathbf{k}')\right] + \sum_{\mathbf{k}''} V(\mathbf{k}''-\mathbf{k}')\left(\psi_{0,\mathbf{k}}(\mathbf{k}') + \psi_{\mathbf{k},\text{scat}}(\mathbf{k}')\right) \\ &\quad = \frac{\mathbf{k}^2}{m}\left(\psi_{0,\mathbf{k}}(\mathbf{k}') + \psi_{\mathbf{k},\text{scat}}(\mathbf{k}')\right) \end{aligned} \tag{A.4}$$

so since $\psi_{0,\mathbf{k}}(\mathbf{k}') = \delta_{\mathbf{k},\mathbf{k}'}$, we obtain

$$\left[\frac{k^2 - k'2}{m}\right]\psi_{\mathbf{k},\text{scat}}(\mathbf{k}') - \sum_{\mathbf{k}''} V(\mathbf{k}''-\mathbf{k}')\psi_{\mathbf{k},\text{scat}}(\mathbf{k}'') = V(\mathbf{k}-\mathbf{k}'). \tag{A.5}$$

The Lippmann-Schwinger equation, Eq. 2.28, was derived in Sect. 2.2.1. The "on-shell" part with $E = k^2/m$ is

$$T(\mathbf{k}, \mathbf{k}') = V(\mathbf{k} - \mathbf{k}') + \frac{1}{\Omega} \sum_{\mathbf{k}''} V(\mathbf{k}' - \mathbf{k}'')((k^2 - k''2)/m + i\delta)^{-1} T(\mathbf{k}, \mathbf{k}', E). \tag{A.6}$$

Comparing Eqs. A.5 and A.6 we see that

$$T(\mathbf{k}, \mathbf{k}') = \left(\frac{k^2 - k'2}{m}\right) \psi_{\mathbf{k},\text{scat}}(\mathbf{k}'). \tag{A.7}$$

This establishes the desired relationship between T and ψ_{scat}.

Finally, I relate the low-momentum behavior of the T-matrix to the scattering length. At $k = 0$, Fourier transforming Eq. A.7 with respect to $\mathbf{k}'$ gives

$$\psi_{\mathbf{k}=0,\text{scat}} = -\frac{mT(0,0)}{4\pi r}. \tag{A.8}$$

In the current notation, the scattering length was defined by $\psi_{\mathbf{k}=0,\text{scat}} = -a/r$, so we see that

$$T(0,0) = 4\pi a/m. \tag{A.9}$$

This is the desired relationship between the scattering length and the T-matrix.

Appendix B
Ward Identities for the RF Spectrum for the Bose-Hubbard Model—Vertex Corrections, Symmetries, and Conservation Laws

B.1 Abstract

I derive a symmetry inherited relationship between the vertex function and Green's function, for rf spectra of the Bose-Hubbard model. This is somewhat non-trivial in this context. We start with Ward identities enforcing this symmetry, and then construct an approximation which satisfies these Ward identities by writing the rf spectra in terms of single particle Green's functions and vertex corrections.

B.2 Introduction and Motivation

We had previously calculated the rf spectra of the Bose-Hubbard model using a Lehmann representation method—this assumed that the rf spectra excites single quasiparticle-quasihole excitations. However, this approach failed to satisfy the symmetry that emerges when the two channels' scattering lengths are equal, due to the neglect of interactions between the excited quasiparticle and quasihole (vertex corrections). Here we derive an approximation which remedies this problem.

The Ward identities/vertex corrections derived are applicable regardless of the starting point of the approximation used to obtain the single particle Green's functions: for example, we may use the Lehmann representation approximation introduced elsewhere, a full mean field Green's function, or an RPA green's function which includes quadratic fluctuations.

B.3 Homogeneous Gas: No Lattice

B.3.1 Setup

We start with the Hamiltonian for our system, which consists of a gas with two hyperfine states $\alpha = 1, 2$. The gas will initially be entirely in the $\alpha = 1$ state.

First we define $\psi_\alpha(\mathbf{r})$ to be the boson annihilation operator at position $\mathbf{r}$ in state α, and notate

$$\psi(\mathbf{r}) = \begin{pmatrix} \psi_1(\mathbf{r}) \\ \psi_2(\mathbf{r}) \end{pmatrix}. \tag{B.1}$$

Then the Hamiltonian is

$$H = H_1 + H_2, \tag{B.2}$$

with H_1 and H_2 defined immediately below.

The single particle Hamiltonian H_1 is given by

$$H_1 = \sum_\alpha \int d^3 r \psi^\dagger(\mathbf{r}) \left(-\sum_\alpha \frac{D_\alpha^2}{2m} + V(\mathbf{r}) \right) \psi(\mathbf{r}) \tag{B.3}$$

where the sum over α runs over the three spatial components and

$$D_\alpha \equiv \frac{\partial}{\partial r_\alpha} - ig\sigma \cdot \mathbf{A}_\alpha(\mathbf{r}, t). \tag{B.4}$$

Note that this covariant derivative D_α is a 2×2 matrix operator. Here σ is the vector of Pauli matrices,

$$\sigma = \begin{pmatrix} \sigma^{(x)} \\ \sigma^{(y)} \\ \sigma^{(z)} \end{pmatrix} \tag{B.5}$$

with

$$\sigma^{(x)} = \begin{pmatrix} 0 & 1 \\ 1 & 0 \end{pmatrix} \tag{B.6}$$

$$\sigma^{(y)} = \begin{pmatrix} 0 & -i \\ i & 0 \end{pmatrix} \tag{B.7}$$

$$\sigma^{(z)} = \begin{pmatrix} 1 & 0 \\ 0 & -1 \end{pmatrix} \tag{B.8}$$

We use the superscript notation for the 3-vector direction determining the Pauli matrix, and we reserve subscripts for the $SU(2)$ indices. In the physical scenario of interest, the gauge field is $\mathbf{A}_\alpha = 0$; however, we include it as our formal derivation of the Ward identities requires it.

The interaction Hamiltonian H_2 is given by

$$H_2 = \sum_{\alpha,\beta} \frac{U_{\alpha,\beta}}{2} \int d^3 r \, \psi_\alpha^\dagger(\mathbf{r}) \psi_\beta^\dagger(\mathbf{r}) \psi_\beta(\mathbf{r}) \psi_\alpha(\mathbf{r}) \tag{B.9}$$

and

$$U_{\alpha,\beta} = \frac{4\pi\hbar^2}{m} a_{\alpha,\beta} \tag{B.10}$$

where $a_{\alpha,\beta}$ is the 3D scattering length.

When the interaction coefficients are equal, that is $U_{11} = U_{12} = U_{21} = U_{22}$, the Hamiltonian is symmetric under a global $SU(2)$ transformation $e^{i\sigma\cdot\Lambda}$ where Λ is an arbitrary 3-vector. As a consequence of this symmetry, the rf spectrum will be a single delta function peaked at the vacuum energy splitting Δ.

We would like to ensure that our approximations satisfy this symmetry requirement (which is connected to conservation laws), and this is the bulk of what is explored in these notes.

B.3.2 Showing Covariance of Kinetic Energy

In order to ensure our approximations satisfy the $SU(2)$ invariance requirement, it will turn out to be necessary to first promote the global $SU(2)$ transformation to a local gauge transformation. The field transformations of interest are

$$\psi(\mathbf{r},t) \to \psi'(\mathbf{r},t) = e^{i\sigma\cdot\Lambda(\mathbf{r},t)}\psi(\mathbf{r},t), \tag{B.11}$$

and in this section we calculate the corresponding transformation of the $SU(2)$ gauge field that ensures the covariant derivative is, in fact, covariant under the following $SU(2)$ gauge transformation.

Rather than working with the general gauge transformation, it will suffice to work with infinitesimal gauge transformations Λ. Then the field transformation law is

$$\psi(\mathbf{r},t) \to \psi'(\mathbf{r},t) = [1 + i\sigma\cdot\Lambda(\mathbf{r},t)]\psi(\mathbf{r},t) + O(\Lambda^2). \tag{B.12}$$

The covariant derivative transforms as

$$D_\alpha \to D'_\alpha = \frac{\partial}{\partial r_\alpha} - ig\sigma\cdot\mathbf{A}'_\alpha(\mathbf{r},t), \tag{B.13}$$

defining $\mathbf{A}'_\alpha$, which we now compute. The covariance of the derivative is determined by requiring

$$D'_\alpha = e^{i\sigma\cdot\Lambda(\mathbf{r},t)} D_\alpha e^{-i\sigma\cdot\Lambda(\mathbf{r},t)}, \tag{B.14}$$

as this implies that

$$[\psi'(\mathbf{r},t)]^\dagger (D'_\alpha)^2 \psi'(\mathbf{r},t) = \psi^\dagger(\mathbf{r},t) D_\alpha^2 \psi(\mathbf{r},t). \tag{B.15}$$

Computing the right hand side of Eq. B.14 gives

$$D'_\alpha = D_\alpha + i[\sigma \cdot \Lambda(\mathbf{r}, t), D_\alpha] \tag{B.16}$$

working here, and henceforth, to the lowest appropriate order in Λ. We compute this commutator by acting with it on a test function:

$$i[\sigma \cdot \Lambda(\mathbf{r}, t), D_\alpha]f(\mathbf{r}, t) = i\left[\sigma \cdot \Lambda(\mathbf{r}, t), \left(\frac{\partial}{\partial r_\alpha} - ig\sigma \cdot \mathbf{A}_\alpha(\mathbf{r}, t)\right)\right]f(\mathbf{r}, t) \tag{B.17}$$

$$= i\left\{\left[\sigma \cdot \Lambda(\mathbf{r}, t), \frac{\partial}{\partial r_\alpha}\right] - ig[\sigma \cdot \Lambda(\mathbf{r}, t), \sigma \cdot \mathbf{A}_\alpha(\mathbf{r}, t)]\right\}f(\mathbf{r}, t) \tag{B.18}$$

The first term is

$$\left[\sigma \cdot \Lambda(\mathbf{r}, t), \frac{\partial}{\partial r_\alpha}\right]f(\mathbf{r}, t) = \left(\sigma \cdot \Lambda(\mathbf{r}, t)\frac{\partial f(\mathbf{r}, t)}{\partial r_\alpha} - \sigma \cdot \frac{\partial(\Lambda(\mathbf{r}, t)f(\mathbf{r}, t))}{\partial r_\alpha}\right) \tag{B.19}$$

$$= -\sigma \cdot \frac{\partial \mathbf{\Lambda}(\mathbf{r}, t)}{\partial r_\alpha}f(\mathbf{r}, t) \tag{B.20}$$

The second term of Eq. B.18 is simplified by noticing that

$$[\sigma \cdot \mathbf{v}, \sigma \cdot \mathbf{w}] = 2i\sigma \cdot (\mathbf{v} \times \mathbf{w}) \tag{B.21}$$

and consequently the commutator in Eq. B.18 is

$$[\sigma \cdot \Lambda(\mathbf{r}, t), \sigma \cdot \mathbf{A}_\alpha(\mathbf{r}, t)] = 2i\sigma \cdot (\Lambda(\mathbf{r}, t) \times \mathbf{A}_\alpha(\mathbf{r}, t)). \tag{B.22}$$

Putting these together, Eq. B.18 yields

$$D'_\alpha = D_\alpha - i\sigma \cdot \left(\frac{\partial \Lambda}{\partial r_\alpha} - 2g\Lambda(\mathbf{r}, t) \times \mathbf{A}_\alpha(\mathbf{r}, t)\right) \tag{B.23}$$

Thus we see that to maintain covariance, upon gauge transforming the atom field ψ by Eq. B.11 the $SU(2)$ gauge field $\mathbf{A}_\alpha$ must transform, for infinitesimal Λ, as

$$\mathbf{A}_\alpha \to \mathbf{A}'_\alpha = \mathbf{A}_\alpha + \frac{1}{g}\frac{\partial \Lambda}{\partial r_\alpha} - 2\Lambda(\mathbf{r}, t) \times \mathbf{A}_\alpha(\mathbf{r}, t). \tag{B.24}$$

B.3.3 Partition Function in Path Integral Representation, Including Sources

To derive the Ward identity, we will introduce the partition function for our system, and consider the consequences of its symmetries. We include source terms in the action, so that we may obtain Green's functions by differentiating the partition function with respect to the sources. In the bosonic coherent state path integral description of the system, using the usual notation for the integrals and integration measure, the partition function Z is defined as

$$Z[J, J^*, \mathbf{W}_\mu] = \int \mathcal{D}\psi \mathcal{D}\mathbf{A}_\alpha e^{-(S[\psi,\psi^*,\mathbf{A}_\alpha]+S_J[J,J^*,\mathbf{W}_\mu;\psi,\psi^*,\mathbf{A}_\alpha])/\hbar}. \tag{B.25}$$

with the (imaginary time) action S defined by

$$S[\psi, \psi^*, \mathbf{A}_\alpha] = \int_0^\beta d\tau \left\{ \int d^3r \left[\psi^\dagger(\mathbf{r}, \tau) \left(\hbar \frac{\partial}{\partial \tau} - \mu \right) \psi(\mathbf{r}, \tau) \right] + H[\psi, \psi^*, \mathbf{A}_\alpha] \right\}. \tag{B.26}$$

Note that $\mu = \begin{pmatrix} \mu_1 & 0 \\ 0 & \mu_2 \end{pmatrix}$ is a 2×2 matrix here. The source action S_J is defined to be

$$S_J[J, J^*, \mathbf{W}_\mu; \psi, \psi^*, \mathbf{A}_\alpha] = \int_0^\beta d\tau \int d^3r \left[J^*(\mathbf{r}, t)\psi(\mathbf{r}, \tau) + \psi^*(\mathbf{r}, \tau)J(\mathbf{r}, \tau) + \sum_\alpha \mathbf{W}_\alpha \cdot \mathbf{A}_\alpha \right]. \tag{B.27}$$

Note that in the path integral, we have allowed $\mathbf{A}_\alpha$ to fluctuate, which it doesn't in the original problem.

B.3.4 Partition Function Under $SU(2)$ Transformations

We would like to know how the partition function transforms under $SU(2)$ gauge transformations. There are three relevant parts: (1) the integration measure $\mathcal{D}\psi\mathcal{D}\mathbf{A}_\alpha$, (2) the system action S, and (3) and the source S_J.

Integration Measure

Atom field integration measure—The integration measure $\mathcal{D}\psi$ is trivially invariant under $SU(2)$ gauge transformations, since the field transformation is unitary. A little more explicitly, note that $\mathcal{D}\psi$ is defined as

$$\mathcal{D}\psi = \prod_{\mathbf{r},\tau} \left[d\psi^*_{\mathbf{r},\tau} d\psi_{\mathbf{r},\tau} \right]. \tag{B.28}$$

This is okay formally, though from a mathematical standpoint it might be more rigorous to define the equivalent measure in Fourier space. Then under the field transformation $\psi'(\mathbf{r}, \tau) = e^{i\sigma\cdot\Lambda(\mathbf{r},\tau)}\psi(\mathbf{r}, \tau)$ we have

$$\mathcal{D}\psi \to \mathcal{D}\psi' = \prod_{\mathbf{r},\tau} \left[d\psi^*_{\mathbf{r},\tau} e^{-i\sigma\cdot\Lambda(\mathbf{r},\tau)} e^{i\sigma\cdot\Lambda(\mathbf{r},\tau)} d\psi_{\mathbf{r},\tau} \right] \tag{B.29}$$

$$= \prod_{\mathbf{r},\tau} \left[d\psi^*_{\mathbf{r},\tau} d\psi_{\mathbf{r},\tau} \right] \tag{B.30}$$

$$= \mathcal{D}\psi, \tag{B.31}$$

showing that the field integration measure $\mathcal{D}\psi$ is invariant under $SU(2)$ transformations.

Gauge field integration measure—We only know and care about the gauge field's transformations under infinitesimal transformations, we will only consider the integration measure transformations to $O(\Lambda)$. Explicitly, the integration measure is

$$\mathcal{D}\mathbf{A}_\alpha = \prod_{\mathbf{r},\tau,\alpha} [dA_\alpha(\mathbf{r},\tau)] \tag{B.32}$$

with $d\boldsymbol{v}$ shorthand for $\prod_j dv_j$ when $\boldsymbol{v}$ is a vector and v_j are the components of $\boldsymbol{v}$. (Again it might mathematically be more proper to work in Fourier space, but this should suffice for our purposes.) Recapitulating Eq. B.24,

$$\mathbf{A}_\alpha \to \mathbf{A}'_\alpha = \mathbf{A}_\alpha + \frac{1}{g}\frac{\partial \Lambda}{\partial r_\alpha} - 2\Lambda(\mathbf{r},t) \times \mathbf{A}_\alpha(\mathbf{r},t), \tag{B.33}$$

Intuitively, $\mathbf{A}_\alpha$ is shifted by the $\frac{1}{g}\frac{\partial\Lambda}{\partial r_\alpha}$ term, and since $\Lambda \times \mathbf{A}_\alpha$ is perpendicular to $\mathbf{A}_\alpha$, it is an infinitesimal rotation of $\mathbf{A}$, and thus the integration measure will be invariant.

We can show this formally. It is helpful to rewrite the transformation of Eq. B.33 with $\mathbf{A}'_\alpha$ an operator acting on $\mathbf{A}_\alpha$. This is found by representing the cross product in components, and yields

$$\mathbf{A}'_\alpha(\mathbf{r},\tau) = \frac{1}{g}\frac{\partial \Lambda(\mathbf{r},\tau)}{\partial r_\alpha} + (1 + 2\mathcal{M})\mathbf{A}_\alpha(\mathbf{r},\tau). \tag{B.34}$$

with

$$\mathcal{M} \equiv \begin{pmatrix} 0 & \Lambda_z & -\Lambda_y \\ -\Lambda_z & 0 & \Lambda_x \\ \Lambda_y & -\Lambda_x & 0 \end{pmatrix}. \tag{B.35}$$

Then, since $\frac{1}{g}\frac{\partial\Lambda(\mathbf{r},\tau)}{\partial r_\alpha}$ is a constant shift for each component integrated over (and we are integrating over all space, and $1 + 2\mathcal{M}$ is a linear transformation, each term in the integration measure of Eq. B.32 transforms as

$$d\mathbf{A}_\alpha(\mathbf{r},\tau) \to d\left[\mathbf{A}'_\alpha(\mathbf{r},\tau)\right] = \frac{1}{|\det(1 + 2\mathcal{M})|} d\mathbf{A}_\alpha(\mathbf{r},\tau). \tag{B.36}$$

The determinant of $(1 + 2\mathcal{M})$ is

$$\det(1 + 2\mathcal{M}) = 1 + 4\Lambda^2 + O(\Lambda^4), \tag{B.37}$$

so to $O(\Lambda)$, the determinant is 1 and

$$d\mathbf{A}_\alpha(\mathbf{r},\tau) \to d\left[\mathbf{A}'_\alpha(\mathbf{r},\tau)\right] = d\mathbf{A}_\alpha(\mathbf{r},\tau). \tag{B.38}$$

Hence, the integration measure for the gauge field $\mathbf{A}_\alpha$ is invariant under $SU(2)$ gauge transformations, as claimed.

System Action

Likewise, the system action $S[\psi, \psi^*, \mathbf{A}_\alpha]$ is invariant under $SU(2)$ gauge transformations. To see this note that the chemical potential term is manifestly invariant, and we showed the kinetic energy is invariant in Sect. B.3.2. All that remains is the interaction energy. Defining $n_\alpha \equiv \psi_\alpha^\dagger \psi_\alpha$ and $n \equiv \sum_\alpha n_\alpha$, the interaction term is

$$\sum_{\alpha,\beta} U_{\alpha,\beta} \psi_\alpha^\dagger \psi_\beta^\dagger \psi_\beta \psi_\alpha = \sum_{\alpha,\beta} U_{\alpha,\beta} \psi_\alpha^\dagger n_\beta \psi_\alpha \tag{B.39}$$

$$= \sum_{\alpha,\beta} U_{\alpha,\beta} n_\alpha (n_\beta - \delta_{\alpha\beta}) \tag{B.40}$$

$$= U\left[\sum_{\alpha,\beta} n_\alpha n_\beta - \sum_\alpha n_\alpha\right] \tag{B.41}$$

$$= Un(n-1). \tag{B.42}$$

Source Action

The source action $S_J[J, J^*, \mathbf{W}_\alpha; \psi, \psi^*, \mathbf{A}_\alpha]$ varies under $SU(2)$ transformations, unlike the system action S and the integration measure $\mathcal{D}\psi\mathcal{D}\mathbf{A}$.

The source action S_J transforms as

$$S_J[J, J^*, \mathbf{W}_\alpha; \psi, \psi^*, \mathbf{A}_\alpha] \to S_J'[J, J^*, \mathbf{W}_\alpha; \psi, \psi^*, \mathbf{A}_\alpha] = S_J + \delta S_J \tag{B.43}$$

with

$$\delta S_J = \int_0^\beta d\tau \int d^3r \Big\{ iJ^*(\mathbf{r}, t)\sigma \cdot \Lambda(\mathbf{r}, \tau)\psi(\mathbf{r}, \tau) - i\psi^*(\mathbf{r}, \tau)\sigma \cdot \Lambda(\mathbf{r}, \tau)J(\mathbf{r}, \tau) \\ + \sum_\alpha \mathbf{W}_\alpha(\mathbf{r}, \tau) \cdot \left[\frac{1}{g}\frac{\partial \Lambda(r, \tau)}{\partial r_\alpha} - 2\Lambda(\mathbf{r}, t) \times \mathbf{A}_\alpha(\mathbf{r}, t)\right]\Big\}, \tag{B.44}$$

which follows immediately from expanding $\psi'(\mathbf{r}, \tau) \doteq [1 + i\sigma \cdot \Lambda(\mathbf{r}, \tau)]\psi(\mathbf{r}, \tau)$ and substituting $\mathbf{A}_\alpha$'s transformation in the definition of S_J in Eq. B.27.

B.3.5 Consequences of $SU(2)$ Invariance of the Partition Function

The partition function is invariant under $SU(2)$ gauge transformations of the fields ψ and $\mathbf{A}_\alpha$. This follows by considering a path integral of an arbitrary functional of the fields $f[\psi, \psi^*, \mathbf{A}_\alpha]$:

$$\mathcal{I} = \int \mathcal{D}\psi \mathcal{D}\mathbf{A}_\alpha f[\psi, \psi^*, \mathbf{A}_\alpha]. \tag{B.45}$$

Consider the behavior under an $SU(2)$ gauge transformation of the integrand, f,

$$\begin{aligned}&\int \mathcal{D}\psi \mathcal{D}\mathbf{A}_\alpha f\left[e^{i\sigma\cdot\Lambda}\psi, \psi^* e^{-i\sigma\cdot\Lambda}, \mathbf{A}_\alpha + \frac{1}{g}\frac{\partial\Lambda}{\partial r_\alpha} - 2\Lambda\times\mathbf{A}_\alpha\right] \\ &= \int \mathcal{D}(e^{-i\sigma\cdot\Lambda}\psi')\mathcal{D}\left[\mathbf{A}'_\alpha - \frac{1}{g}\frac{\partial\Lambda}{\partial r_\alpha} + 2\Lambda\times\mathbf{A}_\alpha\right] f[\psi', (\psi')^*, \mathbf{A}'_\alpha].\end{aligned} \tag{B.46}$$

Now, we can rename the integration variables ψ' to ψ and $\mathbf{A}'_\alpha$ to $\mathbf{A}_\alpha$, and recognize the integration measure as that after a gauge transformation generated by $-\Lambda$, which is invariant under this transformation, by our argument in Sect.B.3.4 . The (suppressed) integration limits are also invariant under the $SU(2)$ transformation since they run over all space. Consequently, Eq. B.46 shows that

$$\int \mathcal{D}\psi \mathcal{D}\mathbf{A}_\alpha f\left[e^{i\sigma\cdot\Lambda}\psi, \psi^* e^{-i\sigma\cdot\Lambda}, \mathbf{A}_\alpha + \frac{1}{g}\frac{\partial\Lambda}{\partial r_\alpha} - 2\Lambda\times\mathbf{A}_\alpha\right] = \int \mathcal{D}\psi \mathcal{D}\mathbf{A}_\alpha f[\psi, \psi^*, \mathbf{A}_\alpha], \tag{B.47}$$

and applying this to the path integral expression for the partition function, shows that the partition function is invariant under $SU(2)$ gauge transformations:

$$Z' = Z. \tag{B.48}$$

As a consequence, certain expectation values are constrained to satisfy a relation with the source fields; we derive this relation now. This relation will follow by computing Z', the partition function after $SU(2)$ gauge transforming the integrand, in a second way by Taylor expanding the transformed partition function Z' for infinitesimal gauge transformations. Using the transformations of the actions and integration measures derived above, we find

$$Z'[J, J^*, \mathbf{W}_\mu] = \int \mathcal{D}\psi \mathcal{D}\mathbf{A}_\alpha \, e^{-(S+S_J+\delta S_J)/\hbar} \tag{B.49}$$

$$= \int \mathcal{D}\psi \mathcal{D}\mathbf{A}_\alpha \, e^{-(S+S_J)/\hbar}\left[1 - \frac{\delta S_J}{\hbar}\right] \tag{B.50}$$

$$= Z\left[1 - \frac{\langle\delta S_J\rangle}{\hbar}\right], \tag{B.51}$$

where we have defined

$$\langle \mathcal{O} \rangle \equiv \frac{1}{Z} \int \mathcal{D}\psi \mathcal{D}\mathbf{A}_\alpha \mathcal{O}. \tag{B.52}$$

Equation B.51 that in order to satisfy the identity of Eq. B.48, we must have

$$\langle \delta S_J \rangle = 0. \tag{B.53}$$

Inserting δS_J from Eq. B.44, this gives

$$\begin{aligned} 0 = \int_0^\beta d\tau \int d^3 r \Big\{ & iJ^*(\mathbf{r}, t)\sigma \cdot \Lambda(\mathbf{r}, \tau)\phi(\mathbf{r}, \tau) - i\phi^*(\mathbf{r}, \tau)\sigma \cdot \Lambda(\mathbf{r}, \tau) J(\mathbf{r}, \tau) \\ & + \sum_\alpha \mathbf{W}_\alpha(\mathbf{r}, \tau) \cdot \left[\frac{1}{g} \frac{\partial \Lambda(\mathbf{r}\tau)}{\partial r_\alpha} - 2\Lambda(\mathbf{r}, t) \times \mathcal{A}_\alpha(\mathbf{r}, t) \right] \Big\} \end{aligned} \tag{B.54}$$

where we have defined

$$\phi(\mathbf{r}, \tau) \equiv \langle \psi(\mathbf{r}, \tau) \rangle, \tag{B.55}$$

$$\mathcal{A}_\alpha(\mathbf{r}, \tau) \equiv \langle \mathbf{A}_\alpha(\mathbf{r}, \tau) \rangle. \tag{B.56}$$

Now, we would like to eliminate the integrals in Eq. B.54, by choosing the $\Lambda(\mathbf{r}, \tau)$ to be delta functions, and to pick out the a'th component of the equation. To facilitate this, we handle differentiating the delta function by integrating by parts and dropping the boundary term:

$$\int_0^\beta d\tau \int d^3 r \, \mathbf{W}_\alpha(\mathbf{r}, \tau) \cdot \left(\frac{\partial \Lambda(\mathbf{r}, \tau)}{\partial r_\alpha} \right) = - \int_0^\beta d\tau \int d^3 r \left(\frac{\partial \mathbf{W}_\alpha(\mathbf{r}, \tau)}{\partial r_\alpha} \right) \cdot \Lambda(\mathbf{r}, \tau). \tag{B.57}$$

Furthermore, because we would like to factor our expression into $\Lambda \cdot (\ldots)$, we rewrite the cross product with the identity

$$\mathbf{W}_\alpha \cdot (\Lambda \times \mathcal{A}_\alpha) = \Lambda \cdot (\mathcal{A}_\alpha \times \mathbf{W}_\alpha). \tag{B.58}$$

Then using $\Lambda(\mathbf{r}, \tau) = \hat{a}\delta(\mathbf{r})\delta(\tau)$, Eq. B.54 is

$$\begin{aligned} 0 = & \, iJ^*(\mathbf{r}, t)\sigma^{(a)}\phi(\mathbf{r}, \tau) - i\phi^*(\mathbf{r}, \tau)\sigma^{(a)} J(\mathbf{r}, \tau) \\ & - \sum_\alpha \left[\frac{1}{g} \frac{\partial W_{\alpha,a}(\mathbf{r}, \tau)}{\partial r_\alpha} + 2(\mathcal{A}_\alpha(\mathbf{r}, t) \times \mathbf{W}_\alpha(\mathbf{r}, \tau))_a \right]. \end{aligned} \tag{B.59}$$

This equation is the consequence of the $SU(2)$ symmetry relating expectation values of observables, ϕ and $\mathcal{A}$, in the presence of sources, J and $\mathbf{W}_\mu$.

B.3.6 Connected Green's Functions and Effective Action

We now turn to using the relation of expectation values implied by Eq. B.59 into a relation between the exact Green's functions and one-particle irreducible vertex function which we need for the rf spectra calculation. First, we need to introduce the connected Green's functions' generating functional and the effective action, which we carry out in this section.

Connected Green's Functions Generating Function

We define F implicitly by $Z = e^{-F/\hbar}$ so that

$$F = -\hbar \log Z. \tag{B.60}$$

Then by the usual arguments, since Z is the generating function of Green's functions—both disconnected and connected—it follows that W is the generating function of connected Green's functions.

It is worth pointing out the simplest example of this:

$$\frac{\delta F}{\delta J(\mathbf{r},\tau)} = -\frac{\hbar}{Z}\frac{\delta Z}{\delta J(\mathbf{r},\tau)} \tag{B.61}$$

$$= -\frac{\hbar}{Z}\int \mathcal{D}\psi\mathcal{D}\mathbf{A}_\alpha\left(-\frac{\psi^*(\mathbf{r},\tau)}{\hbar}\right)e^{-(S+S_J)/\hbar} \tag{B.62}$$

$$= \langle\psi^*(\mathbf{r},\tau)\rangle \tag{B.63}$$

$$= \phi^*(\mathbf{r},\tau), \tag{B.64}$$

where the functional derivative of Z follows immediately from Eqs. B.25 and B.27. Similarly

$$\frac{\delta F}{\delta J^*(\mathbf{r},\tau)} = \phi^*(\mathbf{r},\tau), \tag{B.65}$$

$$\frac{\delta F}{\delta \mathbf{W}_\alpha}(\mathbf{r},\tau) = \mathcal{A}_\alpha(\mathbf{r},\tau). \tag{B.66}$$

Effective Action From the Connected Green's Functions' Generating Function

We define the effective action as the Legendre transform of the connected Green's functions' generating function F. To make this precise, we will write J, J^*, and $\mathbf{W}_\alpha$ as a notational shorthand for

$$J = J(\phi, \phi^*, \mathcal{A}_\alpha) \tag{B.67}$$

$$J^* = J^*(\phi, \phi^*, \mathcal{A}_\alpha) \tag{B.68}$$

$$\mathbf{W}_\alpha = \mathbf{W}_\alpha(\phi, \phi^*, \mathcal{A}_\alpha) \tag{B.69}$$

which are defined by inverting the expressions for ϕ, ϕ^*, and $\mathcal{A}_\alpha$ as a function of the sources. Then the effective action is

$$\Gamma[\phi, \phi^*, \mathcal{A}_\alpha] = F[J, J^*, \mathbf{W}_\mu] - \int d^4\xi \left\{ J^*(\xi)\phi(\xi) + \phi^*(\xi)J(\xi) + \sum_\alpha \mathbf{W}_\alpha(\xi) \cdot \mathcal{A}_\alpha(\xi) \right\} \tag{B.70}$$

and we have introduced the 4-vector notation

$$\xi \equiv (t, r). \tag{B.71}$$

An important relationship obeyed by the effective action is that the functional derivatives of Γ, evaluated at a given average field, give the sources necessary to generate that average field:

$$J(\xi) = -\frac{\delta\Gamma}{\delta\phi^*(\xi)}, \tag{B.72}$$

$$J^*(\xi) = -\frac{\delta\Gamma}{\delta\phi(\xi)}, \tag{B.73}$$

$$\mathcal{W}_\alpha(\xi) = -\frac{\delta\Gamma}{\delta\mathcal{A}_\alpha}(\xi). \tag{B.74}$$

We demonstrate this relationship for J, and the derivations for the other sources follow analogously. From the definition in Eq. B.70, we have

$$\begin{aligned}
\frac{\delta\Gamma}{\delta\phi^*(\xi)} &= \frac{\delta F}{\delta\phi^*(\xi)} - J(\xi) \\
&\quad - \int d^4\xi \left[\frac{\delta J^*(\xi)}{\delta\phi^*(\xi)}\phi(\xi) + \phi^*(\xi)\frac{\delta J(\xi)}{\delta\phi^*(\xi)} + \sum_\alpha \left(\frac{\delta\mathbf{W}_\alpha(\xi)}{\delta\phi^*(\xi)} \right) \cdot \mathcal{A}_\alpha(\xi) \right] \\
&= \int d\xi' \left[\frac{\delta F}{\delta J(\xi')}\frac{\delta J(\xi')}{\delta\phi^*(\xi)} + \frac{\delta F}{\delta J^*(\xi')}\frac{\delta J^*(\xi')}{\delta\phi^*(\xi)} + \sum_\alpha \frac{\delta F}{\delta\mathbf{W}_\alpha(\xi')}\frac{\delta\mathbf{W}_\alpha(\xi')}{\delta\phi^*(\xi)} \right] \\
&\quad - J(\xi) - \int d^4\xi \left[\frac{\delta J^*(\xi)}{\delta\phi^*(\xi)}\phi(\xi) + \phi^*(\xi)\frac{\delta J(\xi)}{\delta\phi^*(\xi)} + \sum_\alpha \left(\frac{\delta\mathbf{W}_\alpha(\xi)}{\delta\phi^*(\xi)} \right) \cdot \mathcal{A}_\alpha(\xi) \right] \\
&= -J(\xi),
\end{aligned} \tag{B.75}$$

where the first step is just the chain rule, and the last step follows from Eqs. B.64 B.65, and B.66. This confirms Eq. B.73, and Eqs. B.74 and B.75 follow analogously.

Effective Action in Terms of One-Particle Irreducible Green's Functions

We introduced the effective action Γ, which is constructed so that its functional derivatives yield the one-particle irreducible response functions (although we used the alternative definition in terms of the Legendre transform above, the connection is shown in, e.g., Negele and Orland). Namely one obtains

$$\begin{aligned}\Gamma = &- \int d^4\xi_1 d^4\xi_2 \phi^*(\xi_2) G^{-1}(\xi_2,\xi_1)\phi(\xi_1) \\ &+ \sum_{a\alpha,\beta,\gamma} \int d^4\xi_1 d^4\xi_2 d^4\xi_3 \phi^*_\alpha(\xi_1)\phi_\beta(\xi_2)\mathbf{A}_{\gamma,a}(\xi_3)\left[\Gamma^\gamma_a(\xi_1,\xi_2,\xi_3)\right]_{\alpha,\beta} \\ &+ O(\phi^4, A^2)\end{aligned} \tag{B.76}$$

where G^{-1} is the operator inverse of the exact one particle Green's function and Γ^α_a is the one-particle irreducible exact vertex operator coupling the gauge field to particle-hole excitations. We conform to the unfortunate notation with two distinct Γ's, following the rest of the world. Note that G is a 2×2 matrix here due to the two spin components. One may obtain the general form of the equation via Taylor expansion of Γ and symmetry considerations; showing the terms are the appropriate one-particle irreducible Green's functions (see Negele and Orland) takes a little more work.

B.3.7 Real Space Ward Identities

To obtain the real space Ward identities, we will translate Eq. B.59 into an equation involving functional derivatives with respect to average fields of the effective action, which will give us the desired relation between the Green's functions and vertex function via Eq. B.76.

Equation B.59, evaluated at ξ_3, becomes

$$\begin{aligned}&-i\frac{\delta\Gamma}{\delta\phi(\xi_3)}\sigma^{(a)}\phi(\xi_3) + i\phi^*(\xi_3)\sigma^{(a)}\frac{\delta\Gamma}{\delta\phi^*(\xi_3)} \\ &= \sum_\alpha \left\{ \frac{1}{g}\left[\frac{\partial}{\partial r_\alpha}\left(\frac{\delta\Gamma}{\delta\mathcal{A}_\alpha(\xi_3)}\right)\right] + 2\left[\mathcal{A}_\alpha(\xi_3) \times \left(\frac{\delta\Gamma}{\delta\mathcal{A}_\alpha(\xi_3)}\right)\right]\right\}_a.\end{aligned} \tag{B.77}$$

Now is also an appropriate time to note that the source field $\mathcal{A}$ is zero for the physical situation—we no longer need additional derivatives with respect to $\mathcal{A}$, so we can do this, yielding

$$-i\frac{\delta\Gamma}{\delta\phi(\xi_3)}\sigma^{(a)}\phi(\xi_3) + i\phi^*(\xi_3)\sigma^{(a)}\frac{\delta\Gamma}{\delta\phi^*(\xi_3)} = \frac{1}{g}\sum_\alpha\left[\frac{\partial}{\partial r_\alpha}\left(\frac{\delta\Gamma}{\delta\mathcal{A}_\alpha(\xi_3)}\right)\right]_a. \tag{B.78}$$

In order to relate the Green's functions to the vertex functions, we want to have second and third derivatives of Γ, so we act on Eq. B.77 with $\delta^2/\left[\delta\phi^*_\alpha(\xi_1)\delta\phi_\beta(\xi_2)\right]$; the equation remains valid since it holds for arbitrary fields ϕ and ϕ^*. This gives

$$\frac{\delta^2}{\delta\phi^*_\alpha(\xi_1)\delta\phi_\beta(\xi_2)}\left\{-i\frac{\delta\Gamma}{\delta\phi(\xi_3)}\sigma^{(a)}\phi(\xi_3)+i\phi^*(\xi_3)\sigma^{(a)}\frac{\delta\Gamma}{\delta\phi^*(\xi_3)}\right\}$$
$$=\frac{1}{g}\sum_\mu\left[\frac{\partial}{\partial r_\mu}\left(\frac{\delta^3\Gamma}{\delta\phi^*_\alpha(\xi_1)\delta\phi_\beta(\xi_2)\delta\mathcal{A}_\mu(\xi_3)}\right)\right]_a \tag{B.79}$$

Evaluating this, retaining only the non-zero terms as implied by Eq. B.76, we obtain

$$-i\frac{\delta^2\Gamma}{\delta\phi^*_\alpha(\xi_1)\delta\phi(\xi_3)}\sigma^{(a)}\frac{\delta\phi(\xi_3)}{\delta\phi_\beta(\xi_2)}+i\frac{\delta\phi^*(\xi_3)}{\delta\phi^*_\alpha(\xi_1)}\sigma^{(a)}\frac{\delta^2\Gamma}{\delta\phi_\beta(\xi_2)\delta\phi^*(\xi_3)}$$
$$=\frac{1}{g}\sum_\mu\left[\frac{\partial}{\partial r_\mu}\left(\frac{\delta^3\Gamma}{\delta\phi^*_\alpha(\xi_1)\delta\phi_\beta(\xi_2)\delta\mathcal{A}_\mu(\xi_3)}\right)\right]_a \tag{B.80}$$

so that

$$\sum_{\nu,\mu}\left\{iG^{-1}_{\alpha,\nu}(\xi_1,\xi_3)\sigma^{(a)}_{\nu,\mu}\delta_{\mu\beta}\delta(\xi_3-\xi_2)-i\sigma^{(a)}_{\mu,\nu}G^{-1}_{\nu,\beta}(\xi_3,\xi_2)\delta_{\mu,\alpha}\delta(\xi_3-\xi_1)\right\}$$
$$=\frac{1}{g}\sum_\mu\frac{\partial}{\partial r_\mu}\left[\Gamma^\mu_a(\xi_1,\xi_2,\xi_3)\right]_{\alpha,\beta}. \tag{B.81}$$

so

$$\sum_{\nu}\left\{iG^{-1}_{\alpha,\nu}(\xi_1,\xi_3)\sigma^{(a)}_{\nu,\beta}\delta(\xi_3-\xi_2)-i\sigma^{(a)}_{\alpha,\nu}G^{-1}_{\nu,\beta}(\xi_3,\xi_2)\delta(\xi_3-\xi_1)\right\}$$
$$=\frac{1}{g}\sum_\mu\frac{\partial}{\partial r_\mu}\left[\Gamma^\mu_a(\xi_1,\xi_2,\xi_3)\right]_{\alpha,\beta}. \tag{B.82}$$

This is the desired real space Ward identity.

In the physical case, we take the sources to zero and consequently the Green's functions become diagonal in the spin space, so that

$$G_{\nu,\mu}(\xi,\xi')=\delta_{\nu,\mu}G_\nu(\xi,\xi'). \tag{B.83}$$

Then Eq. B.82 becomes

$$iG^{-1}_\alpha(\xi_1,\xi_3)\sigma^{(a)}_{\alpha,\beta}\delta(\xi_3-\xi_2)-i\sigma^{(a)}_{\alpha,\beta}G^{-1}_\beta(\xi_3,\xi_2)\delta(\xi_3-\xi_1)$$
$$=\frac{1}{g}\sum_\mu\frac{\partial}{\partial r_\mu}\left[\Gamma^\mu_a(\xi_1,\xi_2,\xi_3)\right]_{\alpha,\beta}. \tag{B.84}$$

B.3.8 Fourier Space Ward Identities

Vector Form

Since the rf probe of interest naturally oscillates at fixed frequency and wavevector, and since our Ward identities decouple into independent algebraic equations in Fourier space, we now Fourier transform Eq. B.85. It first pays to take advantage of the fact we are dealing with a translationally invariant system, so we may use

$$G(\xi,\xi') = G(\xi-\xi') \tag{B.85}$$

and

$$\left[\Gamma_a^\mu(\xi,\xi',\xi'')\right] = \left[\Gamma_a^\mu(\xi-\xi'',\xi'-\xi'')\right]. \tag{B.86}$$

Next we calculate the Fourier transforms for these translationally invariant functions, defining our Fourier conventions in the process. We start with G^{-1}:

$$\begin{aligned}\iint d^4\xi d^4\xi' e^{i(k_1\xi+k_2\xi')}G_\alpha^{-1}(\xi-\xi') &= \int d^4\bar{\xi}\int d^4\xi' e^{i\left(k_1\left(\bar{\xi}+\xi'\right)+k_2\xi'\right)}G_\alpha^{-1}(\bar{\xi})\\ &= \left[\int d^4\xi' e^{i(k_1+k_2)\xi'}\right]\int d^4\bar{\xi}e^{ik_1\bar{\xi}}G_\alpha^{-1}(\bar{\xi})\\ &= (2\pi)^4\delta(k_1+k_2)G_\alpha^{-1}(k_1)\end{aligned} \tag{B.87}$$

where we defined

$$G(k) \equiv \int d\xi e^{ik\xi}G(\xi) \tag{B.88}$$

as our Fourier convention here. Similarly, for $\left[\Gamma_a^\mu\right]$, we find

$$\begin{aligned}&\iiint d^4\xi_1 d^4\xi_2 d^4\xi_3 e^{i(k_1\xi_1+k_2\xi_2+k_3\xi_3)}\left[\Gamma_a^\mu(\xi_1-\xi_3,\xi_2-\xi_3)\right]\\ &= \iiint d^4\bar{\xi}_1 d^4\bar{\xi}_2 d^4\xi_3 e^{i\left[k_1\left(\bar{\xi}_1+\xi_3\right)+k_2\left(\bar{\xi}_2+\xi_3\right)+k_3\xi_3\right]}\left[\Gamma_a^\mu(\xi_1-\xi_3,\xi_2-\xi_3)\right]\\ &= \left[\int d^4\xi_3 e^{i(k_1+k_2+k_3)\xi_3}\right]\iint d^4\bar{\xi}_1 d^4\bar{\xi}_2 e^{i\left(k_1\bar{\xi}_1+k_2\bar{\xi}_2\right)}\left[\Gamma_a^\mu(\bar{\xi}_1,\bar{\xi}_2)\right]\\ &= (2\pi)^4\delta(k_1+k_2+k_3)\left[\Gamma_a^\mu(k_1,-k_2)\right]\end{aligned} \tag{B.89}$$

where we use the Fourier convention

$$\left[\Gamma_a^\mu(k_1,-k_2)\right] = \iint d\xi d\xi' e^{i(k_1\xi-k_2\xi')}\left[\Gamma_a^\mu(\xi_1,\xi_2)\right]. \tag{B.90}$$

With these Fourier transforms, we integrate both sides of Eq. B.84 against $\int\int\int d^4\xi_1 d^4\xi_2 d^4\xi_3 e^{i(k\xi_1 - k'\xi_2 - q\xi_3)}$ (this convention should be somewhat natural when we deal with the rf spectra), which yields

$$\begin{aligned}
&\int\int\int d^4\xi_1 d^4\xi_2 d^4\xi_3 e^{i(k\xi_1 - k'\xi_2 - q\xi_3)} \\
&\quad\times \left\{ iG_\alpha^{-1}(\xi_1,\xi_3)\sigma^{(a)}_{\alpha,\beta}\delta(\xi_3-\xi_2) - i\sigma^{(a)}_{\alpha,\beta}G_\beta^{-1}(\xi_3,\xi_2)\delta(\xi_3-\xi_1)\right\} \\
&= \int\int\int d^4\xi_1 d^4\xi_2 d^4\xi_3 e^{i(k\xi_1 - k'\xi_2 - q\xi_3)} \left\{ \frac{1}{g}\sum_\mu \frac{\partial}{\partial r_{3,\mu}}\left[\Gamma^\mu_a(\xi_1,\xi_2,\xi_3)\right]_{\alpha,\beta}\right\}
\end{aligned} \tag{B.91}$$

$$\begin{aligned}
&\Rightarrow \int\int\int d^4\xi_1 d^4\xi_2 \\
&\quad\times \left\{ e^{i(k\xi_1 - (k'+q)\xi_2)} G_\alpha^{-1}(\xi_1-\xi_2)\sigma^{(a)}_{\alpha,\beta} - e^{i((k-q)\xi_1 - k'\xi_2)}\sigma^{(a)}_{\alpha,\beta}G_\beta^{-1}(\xi_1-\xi_2)\right\} \\
&= \frac{1}{g}\int\int\int d^4\xi_1 d^4\xi_2 d^4\xi_3 e^{i(k\xi_1 - k'\xi_2 - q\xi_3)} \left\{ \sum_\mu q_\mu \left[\Gamma^\mu_a(\xi_1-\xi_3,\xi_2-\xi_3)\right]_{\alpha,\beta}\right\}
\end{aligned} \tag{B.92}$$

So that using our Fourier transform conventions and results, we obtain

$$G_\alpha^{-1}(k)\sigma^{(a)}_{\alpha,\beta} - \sigma^{(a)}_{\alpha,\beta}G_\beta^{-1}(k-q) = \frac{1}{g}\sum_\mu q_\mu\left[\Gamma^\mu_a(k,k-q)\right]_{\alpha,\beta}. \tag{B.93}$$

Equation B.93 is the complete set of Ward identities in Fourier space.

Ward Identities Written into Individual Components

Next, we write out the individual components ($a = x, y, z,\ \alpha = 1, 2$, and $\beta = 1, 2$) of Eq. B.93 to better see their structure and to more easily work with them. These are

$$G_1^{-1}(k) - G_2^{-1}(k-q) = \frac{1}{g}\sum_\mu q_\mu\left[\Gamma^\mu_x(k,k-q)\right]_{12} \tag{B.94}$$

$$G_2^{-1}(k) - G_1^{-1}(k-q) = \frac{1}{g}\sum_\mu q_\mu\left[\Gamma^\mu_x(k,k-q)\right]_{21} \tag{B.95}$$

$$-iG_1^{-1}(k) + iG_2^{-1}(k-q) = \frac{1}{g}\sum_\mu q_\mu\left[\Gamma^\mu_y(k,k-q)\right]_{12} \tag{B.96}$$

$$iG_2^{-1}(k) - iG_1^{-1}(k-q) = \frac{1}{g}\sum_\mu q_\mu\left[\Gamma^\mu_y(k,k-q)\right]_{21} \tag{B.97}$$

$$G_1^{-1}(k) - G_1^{-1}(k-q) = \frac{1}{g}\sum_\mu q_\mu \left[\Gamma_z^\mu(k,k-q)\right]_{11} \tag{B.98}$$

$$-G_2^{-1}(k) + G_2^{-1}(k-q) = \frac{1}{g}\sum_\mu q_\mu \left[\Gamma_z^\mu(k,k-q)\right]_{22}. \tag{B.99}$$

and

$$\begin{aligned}
\sum_\mu q_\mu \left[\Gamma_x^\mu(k,k-q)\right]_{11} &= \sum_\mu q_\mu \left[\Gamma_x^\mu(k,k-q)\right]_{22} = \sum_\mu q_\mu \left[\Gamma_y^\mu(k,k-q)\right]_{11} \\
&= \sum_\mu q_\mu \left[\Gamma_y^\mu(k,k-q)\right]_{22} = \sum_\mu q_\mu \left[\Gamma_z^\mu(k,k-q)\right]_{12} \\
&= \sum_\mu q_\mu \left[\Gamma_z^\mu(k,k-q)\right]_{21} = 0.
\end{aligned} \tag{B.100}$$

This is the complete set of Fourier-space, component-by-component Ward identities.

The important Ward identities for rf spectra are Eqs. B.94, B.95, B.96, and B.97.

Ward Identities Separated into Non-Interacting Vertex Plus Corrections

In calculations it is useful to separate expressions into those which apply to a non-interacting system plus additional terms that come from the interactions. We now carry this out for the Green's functions. First recall the definition of the self energies Σ_α, namely

$$G_\alpha^{-1}(k) = G_{0,\alpha}^{-1}(k) + \Sigma_\alpha(k) \tag{B.101}$$

where $G_{0,\alpha}$ is the component α non-interacting system's Green's function. We will similarly define the non-interacting vertex $\left[\gamma_a^\mu\right]_{\alpha,\beta}$ and interaction correction $\left[\Delta\Gamma_a^\mu\right]_{\alpha,\beta}$: $\Gamma_a^\mu = \gamma_a^\mu + \Delta\Gamma_a^\mu$, defining

$$G_{0,1}^{-1}(k) - G_{0,2}^{-1}(k-q) = \frac{1}{g}\sum_\mu q_\mu \left[\gamma_x^\mu(k,k-q)\right]_{12} \tag{B.102}$$

$$G_{0,2}^{-1}(k) - G_{0,1}^{-1}(k-q) = \frac{1}{g}\sum_\mu q_\mu \left[\gamma_x^\mu(k,k-q)\right]_{21} \tag{B.103}$$

$$-iG_{0,1}^{-1}(k) + iG_{0,2}^{-1}(k-q) = \frac{1}{g}\sum_\mu q_\mu \left[\gamma_y^\mu(k,k-q)\right]_{12} \tag{B.104}$$

$$iG_{0,2}^{-1}(k) - iG_{0,1}^{-1}(k-q) = \frac{1}{g}\sum_\mu q_\mu \left[\gamma_y^\mu(k,k-q)\right]_{21} \tag{B.105}$$

$$G_{0,1}^{-1}(k) - G_{0,1}^{-1}(k-q) = \frac{1}{g}\sum_{\mu} q_\mu \left[\gamma_z^\mu(k,k-q)\right]_{11} \tag{B.106}$$

$$-G_{0,2}^{-1}(k) + G_{0,2}^{-1}(k-q) = \frac{1}{g}\sum_{\mu} q_\mu \left[\gamma_z^\mu(k,k-q)\right]_{22}. \tag{B.107}$$

Equations B.94, B.95, B.96, B.97, B.98 and B.99 then imply

$$\Sigma_1(k) - \Sigma_2(k-q) = \frac{1}{g}\sum_{\mu} q_\mu \left[\Delta\Gamma_x^\mu(k,k-q)\right]_{12} \tag{B.108}$$

$$\Sigma_2(k) - \Sigma_1(k-q) = \frac{1}{g}\sum_{\mu} q_\mu \left[\Delta\Gamma_x^\mu(k,k-q)\right]_{21} \tag{B.109}$$

$$-i\Sigma_1(k) + i\Sigma_2(k-q) = \frac{1}{g}\sum_{\mu} q_\mu \left[\Delta\Gamma_y^\mu(k,k-q)\right]_{12} \tag{B.110}$$

$$i\Sigma_2(k) - i\Sigma_1(k-q) = \frac{1}{g}\sum_{\mu} q_\mu \left[\Delta\Gamma_y^\mu(k,k-q)\right]_{21} \tag{B.111}$$

$$\Sigma_1(k) - \Sigma_1(k-q) = \frac{1}{g}\sum_{\mu} q_\mu \left[\Delta\Gamma_z^\mu(k,k-q)\right]_{11} \tag{B.112}$$

$$-\Sigma_2(k) + \Sigma_2(k-q) = \frac{1}{g}\sum_{\mu} q_\mu \left[\Delta\Gamma_z^\mu(k,k-q)\right]_{22}. \tag{B.113}$$

B.4 Lattice (Single Band)

B.4.1 Setup

We now turn to the case of interacting bosons in a lattice and examine the consequences of $SU(2)$ gauge invariance for the matter-gauge field interaction vertex. The arguments will follow completely analogously to the case without a lattice. I will be a little less pedantic in my presentation here.

The two component Bose-Hubbard Hamiltonian

$$H = H_1 + H_2 \tag{B.114}$$

with

$$H_1 = -t\sum_{\langle i,j\rangle} c_i^\dagger e^{\mathrm{a_{ij}}(t)} c_j + \sum_j c_j^\dagger (V_j - \mu) c_j \tag{B.115}$$

and

$$H_2 = \sum_j \left(\sum_{\alpha,\beta} \frac{U_{\alpha\beta}}{2} c^\dagger_{j,\alpha} c^\dagger_{j,\beta} c_{j,\beta} c_{j,\alpha} \right). \tag{B.116}$$

In these equations, $\sum_{\langle i,j \rangle}$ is a sum over nearest neighbors i and j, V_j is the on-site trapping potential, μ is the chemical potential (both V_j and μ are 2×2 matrices), $U_{\alpha\beta}$ are the interaction parameters, c_j is defined as

$$c_j = \begin{pmatrix} c_{j,1} \\ c_{j,2} \end{pmatrix}. \tag{B.117}$$

with $c_{j,\alpha}$ the Bose annihilation operator for a particle at site j in internal state α, and $\mathbf{a_{ij}}$ a 2×2 matrix corresponding to an externally imposed $SU(2)$ lattice gauge field.

Again, as in the continuum case, a vacuum energy splitting between states $\alpha = 1$ and $\alpha = 2$ may be canonically transformed away. Also, we will ignore the V_j trapping potential, as in the continuum case.

B.4.2 Showing Covariance of Kinetic Energy

The Hamiltonian H in Eq. B.114 is invariant under $SU(2)$ transformations; this follows exactly as in the continuum case for each term except the kinetic term. Imposing invariance on the kinetic term will define the transformation rules of the gauge field. That is, we promote our global $SU(2)$ symmetry $c_j \to c'_j = e^{i\sigma\cdot\Lambda} c_j$ to a local gauge symmetry:

$$c_j(t) \to c'_j(t) = e^{i\sigma\cdot\mathbf{\Lambda_j}(t)} c_j. \tag{B.118}$$

The "covariant derivative" kinetic term,

$$T_{ij} \equiv c^\dagger_i e^{i\sigma\cdot\mathbf{a_{ij}}} c_j \tag{B.119}$$

in the Hamiltonian then transforms as

$$T_{ij} \to T'_{ij} = \left(c^\dagger_i e^{-i\sigma\cdot\mathbf{\Lambda_i}(t)} \right) e^{i\sigma\cdot\mathbf{a_{ij}}}(t) \left(e^{i\sigma\cdot\mathbf{\Lambda_j}(t)} c_j \right). \tag{B.120}$$

We will work only to lowest order in the infinitesimal $SU(2)$ gauge transformations $\mathbf{\Lambda_j}$, so that, using the BCH formula, this transformation is

$$T_{ij} \to T'_{ij} = c^\dagger_i e^{i\sigma\cdot\mathbf{a_{ij}}'} c_j. \tag{B.121}$$

with

$$\sigma \cdot \mathbf{a}'_{\mathbf{ij}} = \sigma \cdot \mathbf{a_{ij}} - \mathbf{(1/2)}[\mathbf{i}\sigma \cdot \mathbf{\Lambda a_{ij}(t)}, \mathbf{i}\sigma \cdot \mathbf{a_{ij}(t)}] + \mathbf{i}\sigma \cdot \delta\mathbf{\Lambda_{ij}}(t) + O(\mathbf{\Lambda}^2) \tag{B.122}$$

where

$$\bar{\mathbf{\Lambda}}_{\mathbf{ij}}(t) \equiv \frac{\mathbf{\Lambda}_{\mathrm{i}}(t) + \mathbf{\Lambda}_{\mathrm{j}}(t)}{2} \tag{B.123}$$

and

$$\delta\mathbf{\Lambda}_{\mathbf{ij}}(t) = \mathbf{\Lambda}_{\mathrm{i}}(t) - \mathbf{\Lambda}_{\mathrm{j}}(t). \tag{B.124}$$

Calculating the commutator via $[\sigma \cdot \mathbf{w}, \sigma \cdot \boldsymbol{v}] = 2i\sigma \cdot (\mathbf{w} \times \boldsymbol{v})$, Eq. B.122 gives, to $O(\mathbf{\Lambda}^2)$,

$$\sigma \cdot \mathbf{a_{ij}}' = \sigma \cdot \mathbf{a_{ij}} + (1/2)\sigma \cdot \left(\bar{\mathbf{\Lambda}}_{\mathbf{ij}}(t) \times \mathbf{a_{ij}}(t)\right) + i\sigma \cdot \delta\mathbf{\Lambda}_{\mathbf{ij}}(t). \tag{B.125}$$

Since the Pauli matrices along with the identity matrix form a complete set we can remove the $\sigma\cdot$, obtaining

$$\mathbf{a_{ij}} \to \mathbf{a_{ij}}' = \mathbf{a_{ij}} + (1/2) \cdot \left(\bar{\mathbf{\Lambda}}_{\mathbf{ij}}(t) \times \mathbf{a_{ij}}(t)\right) + i\delta\mathbf{\Lambda}_{\mathbf{ij}}(t), \tag{B.126}$$

which guarantees gauge covariance. Put precisely, the system is gauge invariant under the simultaneous field transformations of Eq. B.118 and B.126.

B.4.3 Partition Function in Path Integral Representation, Including Sources

The path integral, including sources so we can generate the Green's functions, is, using the usual path integral notational conventions,

$$Z[J_j^*, J_j, \mathbf{w_{ij}}] = \int \mathcal{D}c_j(\tau)\mathcal{D}\mathbf{a_{ij}}(\tau)e^{-(S+S_J)/\hbar} \tag{B.127}$$

where the system sans sources action is

$$S[c_j, c_j^*, \mathbf{a_{ij}}] = \int_0^\beta d\tau \sum_j c_j^\dagger(\tau)\left(\frac{\partial}{\partial\tau} - \mu - \sigma \cdot \phi_j(\tau)\right)c_j(\tau) + \int d\tau H[c_j(\tau), c_j^*(\tau), \mathbf{a_{ij}}(\tau)]. \tag{B.128}$$

We introduce the time-component of the gauge field, ϕ, and note that ϕ and the chemical potential, μ, are 2×2 matrices. The source term is

$$S_J[J_j, J_j^*, \mathbf{w_{ij}}, \mathbf{w_{0j}}; c_j, c_j^*, \mathbf{a_{ij}}, \phi_{\mathbf{j}}] = \int_0^\beta d\tau\left[\sum_j \left(J_j^*(\tau)c_j(\tau) + c_j^*(\tau)J_j(\tau) + \mathbf{w}_{\mathbf{j}}^{(\mathbf{0})}(\tau) \cdot \phi_{\mathbf{j}}(\tau)\right) + \sum_{\langle i,j\rangle} \mathbf{w_{ij}}(\tau) \cdot \mathbf{a_{ij}}(\tau)\right]. \tag{B.129}$$

Note that to maintain covariance, ϕ obeys the transformation law

$$\phi_{\mathbf{j}} \to \phi'_j = \phi_{\mathbf{j}} + i\left(\frac{\partial \mathbf{\Lambda}_{\mathbf{j}}(\tau)}{\partial \tau}\right) - 2i\big(\phi_{\mathbf{j}}(\tau) \times \mathbf{\Lambda}_{\mathbf{j}}(\tau)\big). \tag{B.130}$$

Later, we will find it convenient to use the generating functional (I'll say "free energy") F defined by

$$F = -\hbar \log Z. \tag{B.131}$$

Whereas taking functional derivatives of Z with respect to sources generates the total Green's functions, F generates the *connected* Green's functions.

B.4.4 Action Under $SU(2)$ Transformations

We will soon argue that the partition function is invariant under $SU(2)$ gauge transformations of the integrand, and explore the consequences of this. To facilitate this, we determine how S, S_J, and the integration measure $\mathcal{D}c_j\mathcal{D}\mathbf{a}_{\mathbf{ij}}$ vary under an $SU(2)$ gauge transformation. Either by construction (e.g., for the kinetic terms) or the same arguments as in the non-lattice case, S and the integration measure are invariant under $SU(2)$ transformations of c_j and $\mathbf{a}_{\mathbf{ij}}$. It remains to calculate the behavior of S_J under the gauge transformation of the matter and gauge fields from Eq. B.129. This yields

$$\begin{aligned} S_J &\to S'_J \\ &= \int d\tau \Bigg\{ \sum_j \bigg[J_j^*(\tau) e^{i\sigma\cdot\mathbf{\Lambda}_{\mathbf{j}}(\tau)} c_j(\tau) + c_j^*(\tau) e^{-i\sigma\cdot\mathbf{\Lambda}_{\mathbf{j}}(\tau)} J_j(\tau) + \mathbf{w}_j^0(\tau) \\ &\quad \cdot \bigg(\phi_{\mathbf{j}}(\tau) + i\left(\frac{\partial \mathbf{\Lambda}_{\mathbf{j}}(\tau)}{\partial \tau}\right) - 2i\big(\phi_{\mathbf{j}}(\tau) \times \mathbf{\Lambda}_{\mathbf{j}}(\tau)\big)\bigg)\bigg] + \sum_{\langle i,j\rangle} \mathbf{w}_{\mathbf{ij}}(\tau) \\ &\quad \cdot \bigg(\mathbf{a}_{\mathbf{ij}} + i\delta\mathbf{\Lambda}_{\mathbf{ij}}(\tau) + \frac{1}{2}\big(\bar{\mathbf{\Lambda}}_{\mathbf{ij}}(\tau) \times \mathbf{a}_{\mathbf{ij}}(\tau)\big)\bigg)\Bigg\}. \end{aligned} \tag{B.132}$$

Recognizing that henceforth we may set $\mathbf{a}_{\mathbf{ij}} = 0$ and $\phi_{\mathbf{j}} = 0$, since we won't be further differentiating with respect to these variables, to lowest order in $\mathbf{\Lambda}_{\mathbf{j}}(\tau)$ we have $S_J \to S'_J = S_J + \delta S_J$ with

$$\begin{aligned} \delta S_J = &\int d\tau \Bigg\{ \sum_j \bigg[J_j^*(\tau)\big(i\sigma \cdot \mathbf{\Lambda}_{\mathbf{j}}(\tau)\big)c_j(\tau) + c_*^\dagger(\tau)\big(-i\sigma \cdot \mathbf{\Lambda}_{\mathbf{j}}(\tau)\big)J_j(\tau) + \mathbf{w}_{\mathbf{j}}^{(\mathbf{0})}(\tau) \\ &\cdot \left(i\frac{\partial \mathbf{\Lambda}_{\mathbf{j}}(\tau)}{\partial \tau}\right)\bigg] + \sum_{\langle i,j\rangle} \mathbf{w}_{\mathbf{ij}}(\tau) \cdot \big(i\delta\mathbf{\Lambda}_{\mathbf{ij}}(\tau)\big)\Bigg\}. \end{aligned} \tag{B.133}$$

Massaging $\mathbf{\Lambda}_{\mathbf{j}}$ *to factor in* δS_J—We will soon wish to factor $\mathbf{\Lambda}_{\mathbf{i}}(\tau)$ in the equation. This is to easily evaluate the expressions once we choose it to be a spatial

Kronecker delta and a time Dirac delta function, to obtain simple Ward identities. However, in order to do this, we must massage the term involving $\delta\mathbf{\Lambda_{ij}}$, since this involves both $\mathbf{\Lambda_i}$ and $\mathbf{\Lambda_j}$. This is the analog of the integration by parts carried out in the continuum case. The key to the argument is to recognize that $\sum_i \sum_{j\in \text{n.n of } i} f_{ij} = \sum_j \sum_{i\in \text{n.n of } j} f_{ij}$. Then

$$\sum_{\langle i,j\rangle} \mathbf{w_{ij}} \cdot \delta\mathbf{\Lambda_{ij}} = \sum_{\langle i,j\rangle} \mathbf{w_{ij}} \cdot \left(\mathbf{\Lambda}_\mathrm{i} - \mathbf{\Lambda}_\mathrm{j}\right) \tag{B.134}$$

$$= \sum_i \sum_{j\in \text{n.n of } i} \mathbf{w_{ij}} \cdot \mathbf{\Lambda_i} - \sum_i \sum_{j\in \text{n.n of } i} \mathbf{w_{ij}} \cdot \mathbf{\Lambda_j} \tag{B.135}$$

$$= \sum_i \sum_{j\in \text{n.nofi}} \mathbf{w_{ij}} \cdot \mathbf{\Lambda_i} - \sum_j \sum_{i\in \text{n.nofj}} \mathbf{w}_\mathrm{ji} \cdot \mathbf{\Lambda_i} \tag{B.136}$$

$$= \sum_{\langle i,j\rangle} \left(\mathbf{w_{ij}} - \mathbf{w}_\mathrm{ji}\right) \cdot \mathbf{\Lambda_i}. \tag{B.137}$$

Consequently, using this relation and integrating the time component of the gauge field by parts, Eq. B.133 becomes

$$\delta S_J = \int d\tau \Bigg\{ \sum_i \Bigg[J_i^*(\tau)(i\sigma \cdot \mathbf{\Lambda_i}(\tau))c_i(\tau) + c_i^*(\tau)(-i\sigma \cdot \mathbf{\Lambda_i}(\tau))J_i(\tau) - \frac{\partial \mathbf{w_j^{(0)}}(\tau)}{\partial \tau} \cdot \left(i\mathbf{\Lambda_j}(\tau)\right) \Bigg] + i \sum_{\langle i,j\rangle} \left(\mathbf{w_{ij}}(\tau) - \mathbf{w}_{ji}(\tau)\right) \cdot \mathbf{\Lambda}_i(\tau) \Bigg\}. \tag{B.138}$$

B.4.5 Consequences of $SU(2)$ Invariance of the Partition Function

The partition function Z will turn out to be invariant under $SU(2)$ transformations of its integrand, and this will constrain the field expectation values for a given value of the sources. We now demonstrate these two claims in turn.

Partition Function Z is Invariant Under $SU(2)$ Transformation of Integrand

The partition function Z upon gauge transforming the integrand's fields is

$$Z \to Z' = \int \mathcal{D}c_j(\tau)\mathcal{D}\mathbf{a_{ij}}(\tau) \exp\left[-\left(S[c_j', (c_j^*)', \mathbf{a}_{ij}']\right)/\hbar\right] \tag{B.139}$$

with $c_j' = e^{i\sigma\cdot\mathbf{\Lambda_j}(\tau)}c_j$, $(c_j^*)' = c_j^* e^{-i\sigma\cdot\mathbf{\Lambda_j}(\tau)}$, and $\mathbf{a_{ij}'} = \sigma \cdot \mathbf{a_{ij}} - (1/2)[i\sigma \cdot \bar{\mathbf{\Lambda}}_\mathbf{ij}, i\sigma \cdot \mathbf{a_{ij}}] + i\sigma \cdot \delta\mathbf{\Lambda_{ij}}$. Now, we change integration variables to

$$\tilde{c}_j = e^{-i\sigma\cdot\mathbf{\Lambda_j}(\tau)}c_j, \tag{B.140}$$

$$\tilde{c}_j^* = c_j^* e^{i\sigma\cdot\mathbf{\Lambda_j}(\tau)}, \tag{B.141}$$

$$\tilde{\mathbf{a}}_{\mathbf{ij}} = \sigma\cdot\mathbf{a_{ij}} + (1/2)[i\sigma\cdot\bar{\mathbf{\Lambda}}_{\mathbf{ij}}, i\sigma\cdot\mathbf{a_{ij}}] - i\sigma\cdot\delta\mathbf{\Lambda_{ij}} \tag{B.142}$$

so that

$$Z' = \int \mathcal{D}\tilde{c}_j(\tau)\mathcal{D}\tilde{\mathbf{a}}_{\mathbf{ij}}(\tau)\exp\left[-\left(S[c_j, c_j^*, \mathbf{a_{ij}}]\right)/\hbar\right] \tag{B.143}$$

$$= \int \mathcal{D}c_j(\tau)\mathcal{D}\mathbf{a_{ij}}(\tau)\exp\left[-\left(S[c_j, c_j^*, \mathbf{a_{ij}}]\right)/\hbar\right] \tag{B.144}$$

$$= Z, \tag{B.145}$$

where the last step follows since the integration measure and domain are invariant under the gauge transform associated with $-\mathbf{\Lambda_j}$. Hence, $Z' = Z$ as claimed.

Consequences of Z's Invariance

The Invariance of Z under the integrand's gauge transformation implies $\langle\delta S_J\rangle = 0$ as in the continuum case. Then Eq. B.138 gives

$$0 = \left\langle \int d\tau \left\{ \sum_i \left[J_i^*(\tau)(i\sigma\cdot\mathbf{\Lambda_i}(\tau))c_i(\tau) + c_i^*(\tau)(-i\sigma\cdot\mathbf{\Lambda_i}(\tau))J_i(\tau) - \frac{\partial\mathbf{w_j^{(0)}}(\tau)}{\partial\tau} \cdot (i\mathbf{\Lambda_j}(\tau)) \right] + i\sum_{\langle i,j\rangle}(\mathbf{w_{ij}}(\tau) - \mathbf{w}_{ji})\cdot\mathbf{\Lambda}_i(\tau) \right\} \right\rangle. \tag{B.146}$$

Defining the average

$$b_j(\tau) \equiv \langle c_j(\tau)\rangle, \tag{B.147}$$

and noticing that Eq. B.146 holds for arbitrary $\mathbf{\Lambda_j}(\tau)$, we choose $\mathbf{\Lambda_j}(\tau') = \mathbf{\Lambda}\hat{\boldsymbol{a}}\delta_{j,s}\delta(\tau' - \tau)$ and find

$$J_s^*(\tau)\sigma^{(a)}b_s(\tau) - b_s^*(\tau)\sigma^{(a)}J_s(\tau) = \left[\frac{\partial\mathbf{w_s^{(0)}}(\tau)}{\partial\tau} - \sum_{j\in \text{ n.n. of } s}(\mathbf{w}_{sj}(\tau) - \mathbf{w}_{js}(\tau))\right]_a. \tag{B.148}$$

This is an expression of the relation between sources and average fields, which we will manipulate into the relevant Ward identity.

B.4.6 Effective Action

Legende Transform of F

We define the effective action via the Legendre transformation (and as before, this ensures it is the generating functional of the one particle irreducible Green's function and vertex functions):

$$\begin{aligned}\Gamma[b_j(\tau), b_j^*(\tau), \alpha_{\mathbf{ij}}(\tau)] = W[J_j(\tau), J_j^*(\tau), \mathbf{w}_{\mathbf{ij}}(\tau)] \\ - \int d\tau \Bigg\{ \sum_j \Big[J_j^*(\tau) b_j(\tau) + b_j^*(\tau) J_j(\tau) + \mathbf{w}_j^{(0)} \cdot \alpha_{j,0} \Big] \\ + \sum_{\langle i,j \rangle} \mathbf{w}_{\mathbf{ij}}(\tau) \cdot \alpha_{\mathbf{ij}}(\tau) \Bigg\}\end{aligned} \tag{B.149}$$

where $\alpha_{ij} \equiv \langle \mathbf{a}_{\mathbf{ij}} \rangle$, $\alpha_{j,(0)} \equiv \langle \phi_{\mathbf{j}} \rangle$, and $J_j(\tau)$, $J_j^*(\tau)$ and $\mathbf{w}_{\mathbf{ij}}(\tau)$ are the sources which give rise to average fields $b_j(\tau)$, $b_j^*(\tau)$, and $\alpha_{ij}(\tau)$. Following the arguments in the continuum case, Γ's first functional derivatives are

$$\frac{\delta \Gamma}{\delta b_j(\tau)} = -J_j^*(\tau), \tag{B.150}$$

$$\frac{\delta \Gamma}{\delta b_j^*(\tau)} = -J_j(\tau), \tag{B.151}$$

$$\frac{\delta \Gamma}{\delta \alpha_{\mathbf{ij}}}(\tau) = -\mathbf{w}_{\mathbf{ij}}(\tau). \tag{B.152}$$

Effective Action in Terms of One-Particle Irreducible Functions

As discussed in Negele and Orland, one finds that the effective action is

$$\begin{aligned}\Gamma = & - \int d\tau_1 d\tau_2 \sum_{j_1 j_2} b_{j_1}^\dagger(\tau_1) G^{-1}(j_1, \tau_1; j_2, \tau_2) b_{j_2}(\tau_2) \\ & + \int d\tau_1 d\tau_2 d\tau_3 \sum_a \sum_{\alpha,\beta} \sum_{j_1 j_2 j_3} \sum_{s \in \text{ n.n. of } j_3} b_\alpha^\dagger(j_1, \tau_1) b_\beta(j_2, \tau_2) \alpha_{j_3,s}^{(a)}(\tau_3) \\ & \times \left[\Gamma_{\mathbf{a}}^{j_3,s}(j_1, \tau_1; j_2, \tau_2; j_3, \tau_3) \right]_{\alpha,\beta} \\ & + \int d\tau_1 d\tau_2 d\tau_3 \sum_a \sum_{\alpha,\beta} \sum_{j_1 j_2 j_3} b_\alpha^\dagger(j_1, \tau_1) b_\beta(j_2, \tau_2) \alpha_0^{(a)}(j_3, \tau_3) \\ & \times \left[\Gamma_a^0(j_1, \tau_1; j_2, \tau_2; j_3, \tau_3) \right]_{\alpha,\beta} \\ & + O(b^4) + O(\alpha^2)\end{aligned} \tag{B.153}$$

where G is the 2×2 matrix exact single-particle Green's function and Γ_a^{ij} is the one-particle irreducible exact vertex operator coupling the bond i-j gauge field to particle-hole excitations. We have also changed our notation slightly, with the lattice operators now having the site index in parentheses rather than as a subscript.

B.4.7 Real Space Ward Identities

Equation B.158 will yield the Ward identities between the exact one particle irreducible vertex and the exact Green's functions by taking advantage of Eq. B.153. First we rewrite Eq. B.148 by using Eqs. B.152, and evaluating everything at the position and imaginary time j_3, τ_3:

$$\frac{\delta\Gamma}{\delta b(j_3,\tau_3)}\sigma^{(a)}b(j_3,\tau_3) - b^*(j_3,\tau_3)\sigma^{(a)}\frac{\delta\Gamma}{\delta b^*(j_3,\tau_3)}$$
$$= \frac{\partial}{\partial\tau_3}\left(\frac{\delta\Gamma}{\delta\alpha_0^{(a)}(j_3,\tau_3)}\right) - \sum_{s\in\,\mathrm{n.n.of}\ j_3}\left(\frac{\delta\Gamma}{\delta\alpha_{j_3,s}^{(a)}(\tau_3)} - \frac{\delta\Gamma}{\delta\alpha_{s,j_3}^{(a)}(\tau_3)}\right). \quad \text{(B.154)}$$

In order to turn this equation into one involving the irreducible vertex and Green's functions, we functionally differentiate both sides of the equation twice. This is valid since it holds for arbitrary values of the b's. Then

$$\frac{\delta^2}{\delta b_\alpha^*(j_1,\tau_1)\delta b_\beta(j_2,\tau_2)}\left[\frac{\delta\Gamma}{\delta b(j_3,\tau_3)}\sigma^{(a)}b(j_3,\tau_3) - b^*(j_3,\tau_3)\sigma^{(a)}\frac{\delta\Gamma}{\delta b^*(j_3,\tau_3)}\right]$$
$$= \frac{\delta^2}{\delta b_\alpha^*(j_1,\tau_1)\delta b_\beta(j_2,\tau_2)}\left[\frac{\partial}{\partial\tau_3}\left(\frac{\delta\Gamma}{\delta\alpha_0^{(a)}(j_3,\tau_3)}\right)\right.$$
$$\left. - \sum_{s\in\mathrm{n.n.of}\,j_3}\left(\frac{\delta\Gamma}{\delta\alpha_{j_3,s}^{(a)}(\tau_3)} - \frac{\delta\Gamma}{\delta\alpha_{s,j_3}^{(a)}(\tau_3)}\right)\right] \quad \text{(B.155)}$$

which yields

$$\sum_{\mu,\nu}\left[\frac{\delta^2\Gamma}{\delta b_\alpha^*(j_1,\tau_1)\delta b_\mu(j_3,\tau_3)}\sigma_{\mu,\nu}^{(a)}\frac{\delta b_\nu(j_3,\tau_3)}{\delta b_\beta(j_2,\tau_2)} - \frac{\delta b_\mu^*(j_3,\tau_3)}{\delta b_\alpha^*(j_1,\tau_1)}\sigma_{\mu,\nu}^{(a)}\frac{\delta^2\Gamma}{\delta b_\nu^*(j_3,\tau_3)\delta b_\beta(j_2,\tau_2)}\right]$$
$$= \frac{\partial}{\partial\tau_3}\left(\frac{\delta^3\Gamma}{\delta b_\alpha^*(j_1,\tau_1)\delta b_\beta(j_2,\tau_2)\delta\alpha_0^{(a)}(j_3,\tau_3)}\right)$$
$$- \sum_{s\in\mathrm{n.n.of}\,j_3}\left(\frac{\delta^3\Gamma}{\delta b_\alpha^*(j_1,\tau_1)\delta b_\beta(j_2,\tau_2)\delta\alpha_{j_3,s}^{(a)}(\tau_3)}\right.$$
$$\left. - \frac{\delta^3\Gamma}{\delta b_\alpha^*(j_1,\tau_1)\delta b_\beta(j_2,\tau_2)\delta\alpha_{s,j_3}^{(a)}(\tau_3)}\right) \quad \text{(B.156)}$$

so that finally

$$\sum_{\mu}\left[\frac{\delta^2\Gamma}{\delta b^*_\alpha(j_1,\tau_1)\delta b_\mu(j_3,\tau_3)}\sigma^{(a)}_{\mu,\beta}\delta_{j_2,j_3}\delta(\tau_3-\tau_2)\right.$$
$$\left.-\delta_{j_1,j_3}\delta(\tau_3-\tau_1)\sigma^{(a)}_{\alpha,\mu}\frac{\delta^2\Gamma}{\delta b^*_\mu(j_3,\tau_3)\delta b_\beta(j_2,\tau_2)}\right]$$
$$=\frac{\partial}{\partial\tau_3}\left(\frac{\delta^3\Gamma}{\delta b^*_\alpha(j_1,\tau_1)\delta b_\beta(j_2,\tau_2)\delta\alpha^{(a)}_0(j_3,\tau_3)}\right)$$
$$-\sum_{\mathbf{s}\in \text{n.n.of} j_3}\left(\frac{\delta^3\Gamma}{\delta b^*_\alpha(j_1,\tau_1)\delta b_\beta(j_2,\tau_2)\delta\alpha^{(a)}_{j_3,s}(\tau_3)}\right.$$
$$\left.-\frac{\delta^3\Gamma}{\delta b^*_\alpha(j_1,\tau_1)\delta b_\beta(j_2,\tau_2)\delta\alpha^{(a)}_{\mathbf{s},j_3}(\tau_3)}\right). \tag{B.157}$$

Evaluating the functional derivatives with Eq. B.153 gives the real space Ward identities

$$\sum_{\mu}\left[\delta_{j_1,j_3}\delta(\tau_3-\tau_1)\sigma^{(a)}_{\alpha,\mu}G^{-1}_{\mu,\beta}(j_3,\tau_3;j_2,\tau_2)-G^{-1}_{\alpha,\mu}(j_1,\tau_1;j_3,\tau_3)\sigma^{(a)}_{\mu,\beta}\delta_{j_2,j_3}\delta(\tau_3-\tau_2)\right]$$
$$=\frac{\partial}{\partial\tau_3}\left[\Gamma^0_a(j_1,\tau_1;j_2,\tau_2;j_3,\tau_3)\right]_{\alpha,\beta}$$
$$-\sum_{\mathbf{s}\in\text{n.n.of } j_3}\left(\left[\Gamma^{j_3,s}_a(j_1,\tau_1;j_2,\tau_2;j_3,\tau_3)\right]_{\alpha,\beta}-\left[\Gamma^{\mathbf{s},j_3}_a(j_1,\tau_1;j_2,\tau_2;j_3,\tau_3)\right]_{\alpha,\beta}\right) \tag{B.158}$$

There is a further simplification, which could have been made much sooner, which is that

$$\left[\Gamma^{\mathbf{s}j}\right]=-\left[\Gamma^{j\mathbf{s}}\right]. \tag{B.159}$$

This follows from the fact that $\left[\Gamma^{\mathbf{s}j}_a\right]$ is the variational derivative of the effective action, $\left[\Gamma^{\mathbf{s}j}_a\right]=\delta\Gamma/\delta\alpha^{(a)}_{\mathbf{s},j}$, and that $\alpha^{(a)}_{\mathbf{s},j}=-\alpha^{(a)}_{j,s}$ by definition. Consequently, Eq. B.158 is

$$\sum_{\mu}\left[\delta_{j_1,j_3}\delta(\tau_3-\tau_1)\sigma^{(a)}_{\alpha,\mu}G^{-1}_{\mu,\beta}(j_3,\tau_3;j_2,\tau_2)-G^{-1}_{\alpha,\mu}(j_1,\tau_1;j_3,\tau_3)\sigma^{(a)}_{\mu,\beta}\delta_{j_2,j_3}\delta(\tau_3-\tau_2)\right]$$
$$=\frac{\partial}{\partial\tau_3}\left[\Gamma^0_a(j_1,\tau_1;j_2,\tau_2;j_3,\tau_3)\right]_{\alpha,\beta}-2\sum_{\mathbf{s}\in\text{n.n. of } j_3}\left[\Gamma^{j_3,s}_a(j_1,\tau_1;j_2,\tau_2;j_3,\tau_3)\right]_{\alpha,\beta} \tag{B.160}$$

B.4.8 Fourier Space Ward Identities

As in the continuum case, the translational invariance of our system will lead to simplifications. In particular, we should look at our functions in Fourier space and use the real space translational invariance properties

$$G(\mathbf{s}_1,\tau_1;\mathbf{s}_2,\tau_2) = G(\mathbf{s}_1-\mathbf{s}_2,\tau_1-\tau_2), \tag{B.161}$$

$$\left[\Gamma^0(\mathbf{s}_1,\tau_1;\mathbf{s}_2,\tau_2;\mathbf{s}_3,\tau_3)\right] = \left[\Gamma^0(\mathbf{s}_1-\mathbf{s}_3,\tau_1-\tau_3;\mathbf{s}_2-\mathbf{s}_3,\tau_2-\tau_3)\right], \tag{B.162}$$

and

$$[\Gamma^{r_1,r_2}(\mathbf{s}_1,\tau_1;\mathbf{s}_2,\tau_2;\mathbf{s}_3,\tau_3)] = [\Gamma^{r_1-\mathbf{s}_3,r_2-\mathbf{s}_3}(\mathbf{s}_1-\mathbf{s}_3,\tau_1-\tau_3;\mathbf{s}_2-\mathbf{s}_3,\tau_2-\tau_3)]. \tag{B.163}$$

We Fourier transform each of these equations, yielding

$$\int\int d\tau_1 d\tau_2 \sum_{\mathbf{s}_1,\mathbf{s}_2} e^{i(\mathbf{k}_1\cdot\mathbf{s}_1+\mathbf{k}_2\cdot\mathbf{s}_2+\omega_1\tau_1+\omega_2\tau_2)} G^{-1}(\mathbf{s}_1-\mathbf{s}_2,\tau_1-\tau_2)$$

$$= \int\int d\chi d\tau_2 \sum_{\ell,\mathbf{s}_2} e^{i(\mathbf{k}_1\cdot(\ell+\mathbf{s}_2)+\mathbf{k}_2\cdot\mathbf{s}_2))+\omega_1(\chi+\tau_2)+\omega_2\tau_2)} G^{-1}(\ell,\chi) \tag{B.164}$$

$$= \int d\chi \sum_{\ell} e^{i(\mathbf{k}_1\cdot\ell+\omega_1\chi)} G^{-1}(\ell,\chi) \left[\int d\tau_2 \sum_{\mathbf{s}_2} e^{i((\mathbf{k}_1+\mathbf{k}_2)\cdot\mathbf{s}_2+(\omega_1+\omega_2)\tau_2)}\right] \tag{B.165}$$

$$= 2\pi N\delta(\mathbf{k}_1+\mathbf{k}_2)\delta(\omega_1+\omega_2)\int d\chi \sum_{\ell} e^{i(\mathbf{k}_1\cdot\ell+\omega_1\chi)} G^{-1}(\mathbf{k}_1,\omega_1) \tag{B.166}$$

where we defined

$$G^{-1}(\mathbf{k},\omega) = \int d\chi \sum_{\mathbf{s}} e^{i(\mathbf{k}\cdot\mathbf{s}+\omega\tau)} G(\mathbf{s},\tau). \tag{B.167}$$

For $\left[\Gamma_a^{jj'}\right]$, we similarly obtain

$$\int\int\int d\tau_1 d\tau_2 d\tau_3 \sum_{\mathbf{s}_1,\mathbf{s}_2,\mathbf{s}_3} e^{i(\mathbf{k}_1\cdot\mathbf{s}_1+\mathbf{k}_2\cdot\mathbf{s}_2+\mathbf{k}_3\cdot\mathbf{s}_3+\omega_1\tau_1+\omega_2\tau_2+\omega_3\tau_3)}$$
$$[\Gamma^{r_1-\mathbf{s}_3,r_2-\mathbf{s}_3}(\mathbf{s}_1-\mathbf{s}_3,\tau_1-\tau_3;\mathbf{s}_2-\mathbf{s}_3,\tau_2-\tau_3)]$$
$$= \int\int\int d\chi_1 d\chi_2 d\tau_3 \sum_{\ell_1,\ell_2,\mathbf{s}_3} e^{i(\mathbf{k}_1\cdot(\ell_1+\mathbf{s}_3)+\mathbf{k}_2\cdot(\ell_2+\mathbf{s}_3)+\mathbf{k}_3\cdot\mathbf{s}_3+\omega_1(\chi_1+\tau_3)+\omega_2(\chi_2+\tau_3)+\omega_3\tau_3)}$$
$$\times [\Gamma^{r_1-\mathbf{s}_3,r_2-\mathbf{s}_3}(\ell_1,\chi_1;\ell_2,\chi_2)]$$
$$= \int\int\int d\chi_1 d\chi_2 d\tau_3 \sum_{\ell_1,\ell_2,\mathbf{s}_3} e^{i(\mathbf{k}_1\cdot\ell_1+\mathbf{k}_2\cdot\ell_2+\omega_1\chi_1+\omega_2\chi_2)} e^{i((\mathbf{k}_1+\mathbf{k}_2+\mathbf{k}_3)\cdot\mathbf{s}_3+(\omega_1+\omega_2+\omega_3)\tau_3)}$$
$$\times [\Gamma^{r_1-\mathbf{s}_3,r_2-\mathbf{s}_3}(\ell_1,\chi_1;\ell_2,\chi_2)]. \tag{B.168}$$

Further simplification is possible by considering the superscripts on the Γ and the structure of the vertex term in the Ward identities. In particular, adapting to the notation used in this section, the indices have the form $\sum_{j\in \text{ n.n. of } \mathbf{s}_3} \Gamma^{j,\mathbf{s}_3}$. Consequently, when we shift the summation variables by $\mathbf{s}_3$, the second index vanishes, while the first goes to $j - \mathbf{s}_3$. Furthermore, since the j-summation runs over the nearest neighbors of $\mathbf{s}_3$, the indices $j - \mathbf{s}_3$ in the summation are independent of $\mathbf{s}_3$. Dropping the first index since it is zero, we have

$$\begin{aligned}
&\iiint d\tau_1 d\tau_2 d\tau_3 \sum_{\mathbf{s}_1,\mathbf{s}_2,\mathbf{s}_3} e^{i(\mathbf{k}_1\cdot\mathbf{s}_1+\mathbf{k}_2\cdot\mathbf{s}_2+\mathbf{k}_3\cdot\mathbf{s}_3+\omega_1\tau_1+\omega_2\tau_2+\omega_3\tau_3)} \\
&\left[\Gamma^{\zeta}(\mathbf{s}_1-\mathbf{s}_3,\tau_1-\tau_3;\mathbf{s}_2-\mathbf{s}_3,\tau_2-\tau_3)\right] \\
&= \iiint d\chi_1 d\chi_2 \sum_{\ell_1,\ell_2} e^{i(\mathbf{k}_1\cdot\ell_1+\mathbf{k}_2\cdot\ell_2+\omega_1\chi_1+\omega_2\chi_2)} \left[\Gamma^{\zeta}(\ell_1,\chi_1;\ell_2,\chi_2)\right] \\
&\qquad \times \left[\int d\tau_3 \sum_{\mathbf{s}_3} e^{i((\mathbf{k}_1+\mathbf{k}_2+\mathbf{k}_3)\cdot\mathbf{s}_3+(\omega_1+\omega_2+\omega_3)\tau_3)}\right] \\
&= 2\pi N \delta_{\mathbf{k}_1+\mathbf{k}_2+\mathbf{k}_3}\delta(\omega_1+\omega_2+\omega_3)\left[\Gamma^{\zeta}(\mathbf{k}_1,\omega_1;-\mathbf{k}_2,-\omega_2)\right]. \qquad \text{(B.169)}
\end{aligned}$$

Integrating and summing against

$$\int d\tau_1 \int d\tau_2 \int d\tau_3 \sum_{\mathbf{j}_1,\mathbf{j}_2,\mathbf{j}_3} e^{i(\mathbf{k}\cdot\mathbf{j}_1-\mathbf{k}'\cdot\mathbf{j}_2-\mathbf{q}\cdot\mathbf{j}_3+\omega\tau_1-\omega'\tau_2-\nu\tau_3)},$$

we obtain

$$\begin{aligned}
&\iiint d\tau_1 d\tau_2 d\tau_3 \\
&\quad \times \sum_{\mathbf{j}_1,\mathbf{j}_2,\mathbf{j}_3} e^{i(\mathbf{k}\cdot\mathbf{j}_1-\mathbf{k}'\cdot\mathbf{j}_2-\mathbf{q}\cdot\mathbf{j}_3+\omega\tau_1-\omega'\tau_2-\nu\tau_3)} \sum_{\mu} \delta_{j_1,j_3}\delta(\tau_3-\tau_1)\sigma^{(a)}_{\alpha,\mu}G^{-1}_{\mu,\beta}(j_3,\tau_3;j_2,\tau_2) \\
&\quad - \iiint d\tau_1 d\tau_2 d\tau_3 \sum_{\mathbf{j}_1,\mathbf{j}_2,\mathbf{j}_3} e^{i(\mathbf{k}\cdot\mathbf{j}_1-\mathbf{k}'\cdot\mathbf{j}_2-\mathbf{q}\cdot\mathbf{j}_3+\omega\tau_1-\omega'\tau_2-\nu\tau_3)} \\
&\quad \times \sum_{\mu} G^{-1}_{\alpha,\mu}(j_1,\tau_1;j_3,\tau_3)\sigma^{(a)}_{\mu,\beta}\delta_{j_2,j_3}\delta(\tau_3-\tau_2) \\
&= \iiint d\tau_1 d\tau_2 d\tau_3 \sum_{\mathbf{j}_1,\mathbf{j}_2,\mathbf{j}_3} e^{i(\mathbf{k}\cdot\mathbf{j}_1-\mathbf{k}'\cdot\mathbf{j}_2-\mathbf{q}\cdot\mathbf{j}_3+\omega\tau_1-\omega'\tau_2-\nu\tau_3)} \qquad \text{(B.170)} \\
&\qquad \times \frac{\partial}{\partial\tau_3}\left[\Gamma^{0}_{a}(j_1,\tau_1;j_2,\tau_2;j_3,\tau_3)\right]_{\alpha,\beta} \\
&\qquad - \iiint d\tau_1 d\tau_2 d\tau_3 \sum_{\mathbf{j}_1,\mathbf{j}_2,\mathbf{j}_3} e^{i(\mathbf{k}\cdot\mathbf{j}_1-\mathbf{k}'\cdot\mathbf{j}_2-\mathbf{q}\cdot\mathbf{j}_3+\omega\tau_1-\omega'\tau_2-\nu\tau_3)} \qquad \text{(B.171)}
\end{aligned}$$

$$\times \left[\Gamma_a^{j_3,s}(j_1,\tau_1;j_2,\tau_2;j_3,\tau_3)\right]_{\alpha,\beta} \tag{B.172}$$

Consequently, Fourier transforming Eq. B.160 with the advantage of this translational invariance and also using the fact that the Green's functions are diagonal,

$$G_{\alpha\beta} = \delta_{\alpha,\beta} G_\alpha, \tag{B.173}$$

we obtain

$$\begin{aligned}
&2\pi N\delta(\mathbf{k}-\mathbf{q}-\mathbf{k}',\omega-\nu-\omega')\left\{\sigma^{(a)}_{\alpha,\beta}G_\mu^{-1}(\mathbf{k}-\mathbf{q},\omega-\nu)-G_\alpha^{-1}(\mathbf{k}_1;\omega)\sigma^{(a)}_{\alpha,\beta}\right\}\\
&=(2\pi N\delta_{\mathbf{k}-\mathbf{q}-\mathbf{k}'}\delta(\omega-\nu-\omega'))i\nu\left[\Gamma_a^0(\mathbf{k},\omega;-\mathbf{k}',-\omega')\right]_{\alpha,\beta}\\
&\quad-2\iiint d\tau_1 d\tau_2 d\tau_3\sum_{\mathbf{j}_1,\mathbf{j}_2,\mathbf{j}_3} e^{i(\mathbf{k}\cdot\mathbf{j}_1-\mathbf{k}'\cdot\mathbf{j}_2-\mathbf{q}\cdot\mathbf{j}_3+\omega\tau_1-\omega'\tau_2-\nu\tau_3)}\\
&\quad\times\sum_{s\in\ \text{n.n.of}\ j_3}\left[\Gamma_a^{j_3,s}(j_1,\tau_1;j_2,\tau_2;j_3,\tau_3)\right]_{\alpha,\beta}.
\end{aligned} \tag{B.174}$$

This result may be simplified further. For example, for a cubic lattice, we may exploit translational invariance to write

$$\begin{aligned}
&\iiint d\tau_1 d\tau_2 d\tau_3\sum_{\mathbf{j}_1,\mathbf{j}_2,\mathbf{j}_3} e^{i(\mathbf{k}\cdot\mathbf{j}_1-\mathbf{k}'\cdot\mathbf{j}_2-q\cdot\mathbf{j}_3+\omega\tau_1-\omega'\tau_2-\nu\tau_3)}\\
&\quad\times\sum_{s\in\ \text{n.n.of}\ j_3}\left[\Gamma_a^{j_3,s}(j_1,\tau_1;j_2,\tau_2;j_3,\tau_3)\right]_{\alpha,\beta}\\
&=\iiint d\tau_1 d\tau_2 d\tau_3\sum_{\mathbf{j}_1,\mathbf{j}_2,\mathbf{j}_3} e^{i(\mathbf{k}\cdot\mathbf{j}_1-\mathbf{k}'\cdot\mathbf{j}_2-q\cdot\mathbf{j}_3+\omega\tau_1-\omega'\tau_2-\nu\tau_3)}\\
&\quad\times\sum_{\zeta=\pm x,\pm y,\pm z}\left[\Gamma_a^{0,\zeta}(j_1-j_3,\tau_1-\tau_3;j_2-j_3,\tau_2-\tau_3)\right]_{\alpha,\beta}\\
&=\iiint d\tau_1 d\tau_2 d\tau_3\sum_{\mathbf{j}_1,\mathbf{j}_2,\mathbf{j}_3} e^{i(\mathbf{k}\cdot\mathbf{j}_1-\mathbf{k}'\cdot\mathbf{j}_2-q\cdot\mathbf{j}_3+\omega\tau_1-\omega'\tau_2-\nu\tau_3)}\\
&\quad\times\sum_{\zeta=x,y,Z}\Big[\Gamma_a^{0,\zeta}(j_1-j_3,\tau_1-\tau_3;j_2-j_3,\tau_2-\tau_3)\\
&\quad+\Gamma_a^{0,-\zeta}(j_1-j_3,\tau_1-\tau_3;j_2-j_3,\tau_2-\tau_3)\Big]_{\alpha,\beta}.
\end{aligned} \tag{B.175}$$

Then the final simplification comes from using translational invariance to write

$$\left[\Gamma_a^{0,-\zeta}(j_1,\tau_1;j_2,\tau_2)\right]_{\alpha,\beta}=\left[\Gamma_a^{\zeta,0}(j_1+\zeta,\tau_1;j_2+\zeta,\tau_2)\right]_{\alpha,\beta} \tag{B.176}$$

and the bond-reversal anti-symmetry of $\mathbf{a_{ij}}$ exploited earlier to write

$$\left[\Gamma_a^{\zeta,0}(j_1+\zeta,\tau_1;j_2+\zeta,\tau_2)\right]_{\alpha,\beta} = -\left[\Gamma_a^{0,\zeta}(j_1+\zeta,\tau_1;j_2+\zeta,\tau_2)\right]_{\alpha,\beta}. \tag{B.177}$$

Using this in Eq. B.175 yields

$$\begin{aligned}
&\iiint d\tau_1 d\tau_2 d\tau_3 \sum_{\mathbf{j_1,j_2,j_3}} e^{i(\mathbf{k}\cdot\mathbf{j_1}-\mathbf{k}'\cdot\mathbf{j_2}-\mathbf{q}\cdot\mathbf{j_3}+\omega\tau_1-\omega'\tau_2-\nu\tau_3)} \\
&\quad\times \sum_{s\in \text{ n.n. of } j_3} \left[\Gamma_a^{j_3,s}(j_1,\tau_1;j_2,\tau_2;j_3,\tau_3)\right]_{\alpha,\beta} \\
&= \iiint d\tau_1 d\tau_2 d\tau_3 \sum_{\mathbf{j_1,j_2,j_3}} e^{i(\mathbf{k}\cdot\mathbf{j_1}-\mathbf{k}'\cdot\mathbf{j_2}-\mathbf{q}\cdot\mathbf{j_3}+\omega\tau_1-\omega'\tau_2-\nu\tau_3)} \\
&\quad\times \sum_{\zeta=x,y,z} \left[\Gamma_a^{0,\zeta}(j_1,\tau_1;j_2,\tau_2) - \Gamma_a^{0,\zeta}(j_1+\zeta,\tau_1;j_2+\zeta,\tau_2)\right]_{\alpha,\beta}.
\end{aligned} \tag{B.178}$$

So finally, substituting this vertex Fourier transform, and the formulas for vertex Fourier transforms from earlier, into Eq. B.175, we obtain

$$\begin{aligned}
\sigma_{\alpha,\beta}^{(a)} G_\mu^{-1}(\mathbf{k}-\mathbf{q},\omega-\nu) - G_\alpha^{-1}(\mathbf{k};\omega)\sigma_{\alpha,\beta}^{(a)} &= i\nu\left[\Gamma_a^0(\mathbf{k},\omega;\mathbf{k}-\mathbf{q},\omega-\nu)\right]_{\alpha,\beta} \\
&\quad - 2\sum_{\zeta=\hat{x},\hat{y},\hat{z}} \left(1-e^{i\mathbf{q}\cdot\zeta}\right)\left[\Gamma_a^{0,\zeta}(\mathbf{k},\omega;-\mathbf{k}',-\omega')\right]_{\alpha,\beta}
\end{aligned} \tag{B.179}$$

This is final momentum-space expression of the Ward identity for the $SU(2)$ symmetric Bose-Hubbard model. Note that as $\mathbf{q}\to 0$, it reduces to the expression obtained in the continuum—the factor of two difference is because in the lattice case we counted each bond twice. We could also write these out component-by-component, analogous to Eqs. B.94, B.95, B.96, and B.97. The form is identical, replacing the spatial q_μ's by the $2(1-e^{iq_\mu|\zeta})$.

B.5 Rf Spectra

Given Eqs. B.93 and B.179, we would like to examine the consequences for the rf spectra. Although my main goal is to determine the rf spectra for the Bose-Hubbard model (which will be explained elsewhere), here we would like to check that things work for a dilute gas.

B.5.1 Rf Spectra in Terms of Exact Single Particle Green's Functions and Interaction Vertex

Pethick and Stoof's PRA, Stoof's notes, and Stoof's book give the rf spectra $I(\omega)$ as the analytic continuation of $I(i\nu_n)$ (where ν_n is a Matsubara frequency), with

$$I(i\nu_n) \propto \sum_{\omega_n} G_2(i\omega_n)G_1(i\omega_n - i\nu_n) \times \left[\gamma_x^0(i\omega_n, i\omega_n - i\nu_n) + \Delta\Gamma_x^0(i\omega_n, i\omega_n - i\nu_n)\right]_{12} \tag{B.180}$$

This is a standard result for the two-particle response functions in terms of Green's functions and vertices.

We will use our results for the Green's functions G_α and vertex corrections $\Delta\Gamma$ to calculate this for the dilute gas now.

B.5.2 Non-Interacting Gas

In the non-interacting gas, the self energy is zero. Also, we may canonically transformed away the vacuum energy difference between states 1 and 2. Then defining the dispersion

$$\epsilon_{\mathbf{k}} = \frac{\hbar^2 k^2}{2m}, \tag{B.181}$$

one finds the Green's functions

$$G_\alpha(i\omega_n) = \frac{1}{i\omega_n - \epsilon_{\mathbf{k}}}. \tag{B.182}$$

Our Ward identities imply that the vertex corrections are zero since the self energies are zero and the system is $SU(2)$ symmetric. Consequently, Eq. B.180 gives the rf spectra to be

$$I(i\nu_n) \propto \sum_{\omega_n} \frac{1}{i\omega_n - \epsilon_{\mathbf{k}}} \frac{1}{i\omega_n - i\nu_n - \epsilon_{\mathbf{k}}}. \tag{B.183}$$

Working at zero temperature, so that the sums are contour integrals, we have

$$I(i\nu_n) \propto \frac{1}{i\nu_n}. \tag{B.184}$$

I have taken the particle pole (from G_2) to be in the upper half of the complex plane and the hole pole (from G_1) to be in the lower complex plane.

As expected, we have an un-shifted delta function peak due to the $SU(2)$ symmetry of the problem.

B.5.3 Dilute Gas: Condensed and Non-Condensed

Condensed gas—In the dilute, fully condensed gas, the self energies are

$$\Sigma_2 = \Sigma_1 = \frac{4\pi\hbar^2 a}{m}\rho, \tag{B.185}$$

where ρ is the particle density. Again, we have $SU(2)$ symmetry and the vertex corrections are zero, so the rf spectrum is an unshifted delta function

$$I(i\nu_n) \propto \frac{1}{i\nu_n} \tag{B.186}$$

identical to the non-interacting gas result, Eq. B.184.

Thermal, non-condensed gas—In the thermal gas, the self energies are

$$\Sigma_2 = \frac{8\pi\hbar^2 a_{12}}{m}\rho \tag{B.187}$$

and

$$\Sigma_1 = \frac{4\pi\hbar^2 a_{11}}{m}\rho, \tag{B.188}$$

where the factor of 2 in Σ_2 comes from the fact that both direct and exchange terms contribute since states 1 and 2 are distinguishable, and that all the particle density ρ is in state 1. Then at the $SU(2)$ symmetric $a_{12} = a_{11}$ point the rf spectra neglecting vertex corrections would be a delta function shifted by $\frac{4\pi\hbar^2 a}{m}\rho$. This is important point raised by Pethick and Stoof regarding the "factor of two" anomaly of a naive mean-field result.

Now I show that going beyond this by including vertex corrections corrects this result to give an unshifted peak at the $SU(2)$ symmetric point. Just for fun, let's be somewhat more general see what rf spectra our vertex corrections derived from the Ward identity give for arbitrary values of a_{12} and a_{11}, even though the Ward identity is only valid for the $SU(2)$ symmetry point with $a_{12} = a_{11}$. We take the vertex correction from Eq. B.108, noting that only the ν component of q is non-zero in an rf spectra experiment. For convenience, we also set the dispersions of each state to zero, since they are the same for the two states. Then the spectrum from Eq. B.183 is

$$I(i\nu_n) \propto \sum_{\omega_n} G_2(i\omega_n)G_1(i\omega_n - i\nu_n)\left(1 + \frac{\Sigma_1 - \Sigma_2}{i\nu_n}\right) \tag{B.189}$$

giving

$$I(i\nu_n) \propto \sum_{\omega_n} \frac{1}{i\omega_n - \Sigma_2} \frac{1}{i\omega_n - i\nu_n - \Sigma_1}\left(1 + \frac{\Sigma_1 - \Sigma_2}{i\nu_n}\right) \tag{B.190}$$

$$= \frac{1}{i\nu_n - (\Sigma_2 - \Sigma_1)}\left(1 + \frac{\Sigma_1 - \Sigma_2}{i\nu_n}\right) \tag{B.191}$$

$$= \frac{1}{i\nu_n}.$$

This is, as it was derived to be, correct in the $SU(2)$ symmetric case, but wrong everywhere else.

A second option is to *literally* use the vertex correction at the $SU(2)$ symmetric point, $\Delta\left[\Gamma_x^0\right]_{12} = -4\pi\hbar^2 a\rho/m$, to which it evaluates there, rather than the symbolic expression. Then the rf spectra is

$$I(i\nu_n) \propto \sum_{\omega_n} \frac{1}{i\omega_n - \Sigma_2} \frac{1}{i\omega_n - i\nu_n - \Sigma_1} \left(1 + \frac{\Sigma_1}{i\nu_n}\right) \tag{B.192}$$

$$= \frac{1}{i\nu_n - (\Sigma_2 - \Sigma_1)} \left(\frac{i\nu_n - \Sigma_1}{i\nu_n}\right) \tag{B.193}$$

but this doesn't help.

To summarize, while we see the derived Ward identities indeed reproduce the appropriate response function when there is $SU(2)$ symmetry, this is a trivial result. Moreover, it is not obvious how to extend this to work when the symmetry is broken, even slightly.

Note: I relied heavily on Dickerscheid, Stoof, et al.'s notes for the fermion rf spectra to derive these Ward identities.

Biographical Sketch

Kaden Hazzard was born 11 January 1982; that night, he started things off right by breaking the window in his parents' car (the subzero temperatures may have also contributed). After graduating high school in 2000, he pursued B.S. degrees in physics, mathematics, and a minor in chemistry at the Ohio State University, leaving in August 2004 for graduate study at Cornell. (Later, he decided it would be a good idea to fill out the paperwork for his Ohio State degrees and was officially awarded his degree in 2008.) He obtained his master's degree in theoretical physics from Cornell University in 2009, and will complete his PhD in May 2010. In June 2010, he began a postdoctoral research fellowship at JILA and the University of Colorado-Boulder working and collaborating with Ana Maria Rey, Victor Gurarie, Michael Hermele, Leo Radzihovsky, and the other faculty there.

Curriculum Vitae

Kaden R. A. Hazzard

University of Colorado
JILA and Department of Physics
JILA S305
Boulder CO 80302
Tel.: +1-60-72208211
Fax: +1 60 72556428
e-mail: kaden.hazzard@colorado.edu
Web:http://jila.colorado.edu/~kaha1718

Education

Cornell University
Degree: M.S. 2009, Ph.D. May 2010
Thesis advisor: Erich J. Mueller
Thesis topic: Quantum phase transitions in cold atoms and low temperature solids

The Ohio State University
Degree: August 2004, B.S. coursework completed with honors in liberal arts
Double major: physics, mathematics. Also, all chemistry minor courses

University of New South Wales. Sydney, Australia
June–December 2003 Study abroad/General education classes

Major Honors and Awards

- GAANN Fellow, 2006–2007
- Special Cornell Graduate Fellowship, 2004–2006
- Hertz Fellowship Finalist, 2004–2005
- Phi Beta Kappa, elected to 2004
- Semiconductor Research Corporation Education Alliance Undergraduate Research Assistant Award: $12 K/year, 2000–2004

Research Interests

I am interested in novel, especially quantum, many-body phenomena. Recently, I have largely focused on ultracold atoms, theoretically exploring experimental techniques to generate and probe strongly correlated states. What excites me most is condensed matter and materials physics: ultracold atoms offer a unique and highly favorable setting to address important questions in those fields. I try to focus my efforts at this interface. I am presently connecting physics of quantum criticality to its manifestations in cold atom experiments. As an undergraduate, I experimented on strongly correlated electron systems and calculated materials' properties via ab initio methods.

Relevant Employment History

June 2010–present. University of Colorado, Boulder and JILA
Postdoctoral Research Fellow with Ana Maria Rey, Victor Gurarie, Michael Hermele: mainly cold atoms theory.

Nov 2004–June 2010. Laboratory of Atomic and Solid-State Physics, Cornell University
Graduate Research Assistant with Erich Mueller: mainly cold atoms theory.

July–December 2003. University of New South Wales
Research Assistant with Alex Hamilton: quantum computing.

June–July 2003. Materials Computation Center, UIUC
Visiting Research Scientist with Jeongnim Kim: materials physics and simulation

June–September 2002. Los Alamos National Laboratories
Research Assistant with Joe D. Thompson: strongly correlated electron experiments

May 2000–August 2004. The Ohio State University Department of Physics
Research Assistant with John W. Wilkins: mainly materials physics and simulations

1997–2000. Coshocton Environmental Testing
Laboratory Technician

Other Education, Research Output, Service, and Positions of Responsibility

I have attended eleven workshops and summer schools, not including conferences. I have organized departmental seminars, chaired several sessions at international conferences, refereed for the New Journal of Physics since 2010, and performed other service activities. I am also an active member of several physics societies.

I have twelve papers published in or submitted to leading physics journals, mostly as first author, and five other refereed papers. Recently, these have focused on the field of cold atoms, with particular attention to areas where experiments can make an impact on condensed matter physics. In the past I have also worked on low-temperature fluids and solids, strongly correlated electron systems, computational physics, and materials science. I have given seventeen invited talks at international conferences (DAMOP) and institutional seminars and colloquia. I have given twenty-two contributed talks and posters. I have taught or graded seven graduate and undergraduate courses and served as a private tutor.

GPSR Compliance

The European Union's (EU) General Product Safety Regulation (GPSR) is a set of rules that requires consumer products to be safe and our obligations to ensure this.

If you have any concerns about our products, you can contact us on ProductSafety@springernature.com

In case Publisher is established outside the EU, the EU authorized representative is:

Springer Nature Customer Service Center GmbH
Europaplatz 3
69115 Heidelberg, Germany

Batch number: 10614744

Printed by Printforce, the Netherlands